I0469981

PHYSICS
on a Shoestring

PHYSICS
on a Shoestring

BINANDA C. BARKAKATY

Copyright © 2011 by Binanda C. Barkakaty.

Library of Congress Control Number:		2011912970
ISBN:	Hardcover	978-1-4653-0293-9
	Softcover	978-1-4653-0292-2
	Ebook	978-1-4653-0294-6

All rights reserved. No part of this book may be reproduced or transmitted in any form or by any means, electronic or mechanical, including photocopying, recording, or by any information storage and retrieval system, without permission in writing from the copyright owner.

This book was printed in the United States of America.

To order additional copies of this book, contact:
Xlibris Corporation
0-800-644-6988
www.xlibrispublishing.co.uk
Orders@xlibrispublishing.co.uk
301911

Contents

PART I: Physics for Early Learners

PART II: Physics for Upper Stages

This book is dedicated to

Puspa, Santana, Biraj, Mark, Cade, and Bailey

Foreword

This book aims to provide a comprehensive compilation of major concepts in physics for students of schools and colleges. It is by no means an exhaustive text book covering all the topics offered by different examination boards. However, this is an attempt to explain each major topic with a concept mapping showing the links between the key words and phrases on that topic followed by a short and simple explanation. In some sections, learning zones are added for easy access to the main formulae. My sincere hope is that students will be able to learn the concepts quickly through the mind maps and be able to enjoy learning physics.

This book has its origin in some of my tutorial sessions with students during vacation times preparing them for their examination. I must therefore offer my grateful indebtedness to many authors of physics text and various student help books, which provided the backdrop for this book. Inspired by the students' feedback and their examination results, I have decided to put the work in a book form for the benefit of all students across the globe.

Helping students in their early secondary stage through young adulthood in making the grade in physics is the main thrust of this book.

Binanda C. Barkakaty

PART I

Physics for Early Learners

Section 1

Force and Movement

Concept Mapping

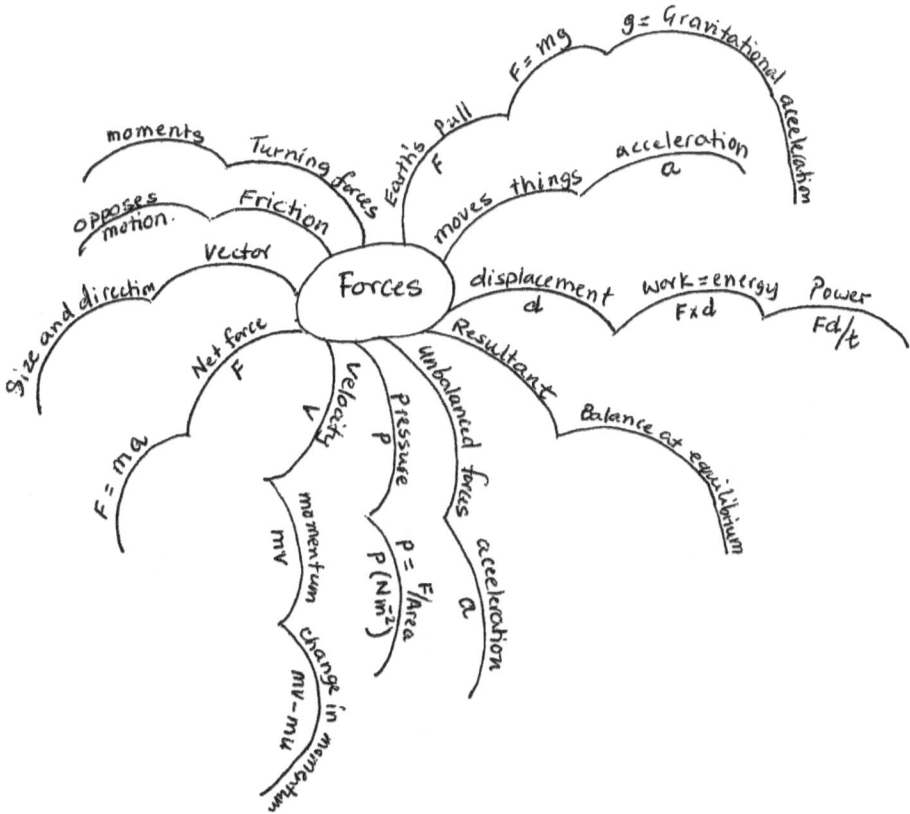

Force is measured in newton (N).
Force has size and direction. So, force is a vector quantity.
Force can be drawn as a line with an arrow using a scale.
Force moves, turns, and changes directions of objects.

Force accelerates and decelerates an object.

Unbalanced forces cause acceleration and deceleration of a body.

For a body moving with a constant speed, there is no acceleration; hence there is no resultant force.

Frictional resistance is a force which opposes motion.

When force is applied on a surface, pressure is produced.

Pressure = force / area, measured in N/M^2 or NM^{-2}.

Pressure is also expressed in pascals (Pa).

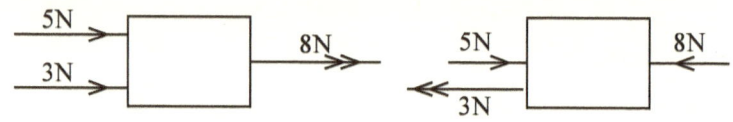

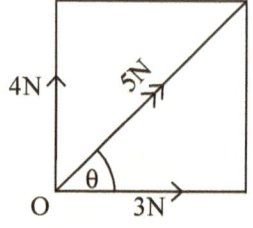

Two forces 3N and 4N act at O, at right angles.
Apply Pythagoras.
Resultant = $\sqrt{(4^2 + 3^2)}$
= 5N
Direction: $\tan\theta = 4/3$
$\theta = \tan^{-1}(4/3)$
$\theta = 53.1°$

The resultant force has a magnitude of 5N.
It is acting at an angle of 53.1° to the horizontal.

Turning forces

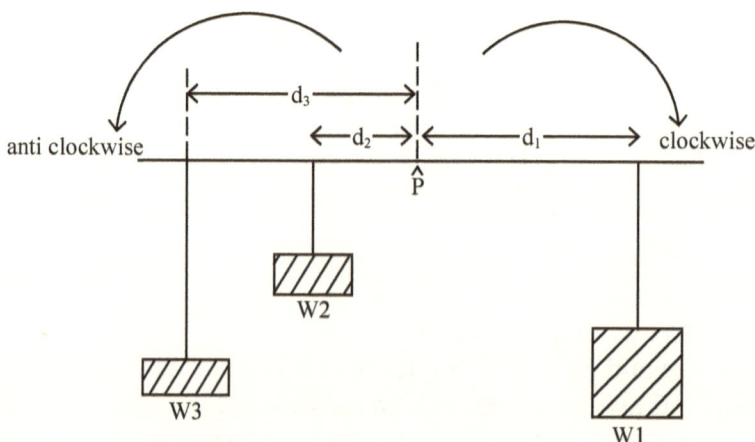

The weights on either side of the pivot P cause a rotational effect on the horizontal ruler. At equilibrium, the sum of the clockwise moments is equal to the sum of the anticlockwise moments.
Therefore, $W_1d_1 = W_2d_2 + W_3d_3$.
Remember, moments have the unit of force × distance.
This gives the unit of moments as newton metre (Nm) or joule (J).

Resistive forces

Friction always opposes motion of an object.
Friction is a force.
A moving car is opposed by friction on the road and the air resistance, which we call drag.

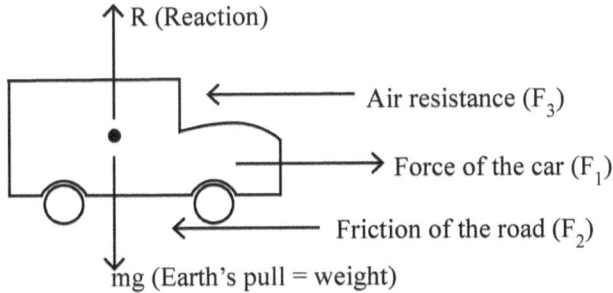

If the car has acceleration, the resultant or unbalanced force becomes

$$F_1-(F_2 + F_3).$$

This net force causes the car to accelerate, therefore,

$$F_1-(F_2 + F_3) = ma,$$

where m is the mass of the car and a is the acceleration.

When the car moves at a constant speed,

$$F_1-(F_2 + F_3) = 0.$$

Then the net force is zero, and hence there is no acceleration of the car.

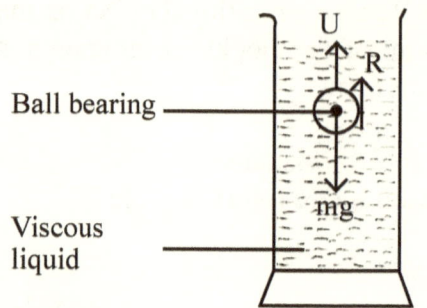

The ball bearing moves vertically downwards at a constant speed when

$$Mg = U + R.$$

The ball bearing is opposed by the resistance of the viscous liquid (drag = R) and upthrust (U) due to the displacement of the liquid.

The constant velocity reached by the ball bearing under these conditions is called the terminal velocity.

Similar situation occurs when a parachutist falls downwards under the action of the gravitational pull.

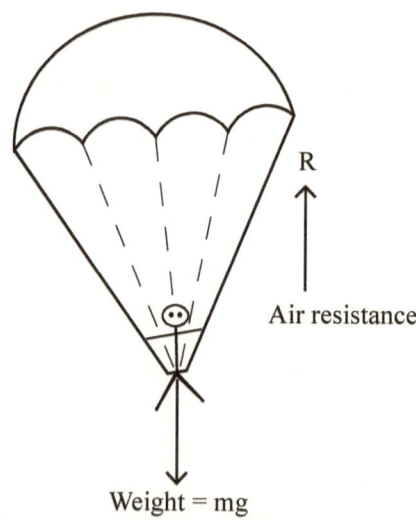

Air resistance

Weight = mg

Had there been no air resistance, all falling bodies would accelerate with the same rate of gravitational field strength g = 9.8 m/s².

As the parachutist accelerates downwards, the air resistance increases. As a result, after a while, the weight becomes equal to the air resistance, and the upward and downward forces cancel out. The parachutist then reaches the terminal velocity with which the parachutist descends.

Let us learn the three laws which govern the movement of a body.

First Law: Without an unbalanced force, nothing moves. There is no change in velocity or change in direction of motion.

Second Law: An unbalanced force produces acceleration. The greater the unbalanced force, the greater the acceleration. Net force = mass × acceleration.

Third Law: If a force is applied on a body, the body then exerts equal and opposite force, along the same line of action.

All masses attract one another. We call it gravitational force. Mass is a scalar quantity. It has only size. The unit of mass is kg. Earth's gravitational field strength is the force on a mass of 1 kg toward the centre of the earth.

This is written as $g = 9.8$ N/kg or 9.8 m/s^2.

This is also known as acceleration due to gravity. The weight of an object is the gravitational pull acting on it.
Therefore,

Weight = mass × acceleration due to gravity, or,
$W = mg$.

The value of mass does not change from place to place, but weight changes due to the changes in the value of g in different places.

On the surface of the earth, $g = 9.8$ N/kg.
On the surface of the moon, $g = 1.6$ N/kg.
On earth, a 10 kg mass weighs $10 × 9.8 = 98$ N.
On the moon, the weight of a 10 kg mass $= 10 × 1.6 = 16$ N.

Learning Zone

Gravitational field strength g = earth's pull on 1 kg mass (N/kg).
Momentum = mass × velocity (kg m/s).
Pressure = force/area (N/m^2 or pascal).
Work done = force × distance moved (Nm or J).
Power = work or energy/time (J/s).
Weight = mass × gravitational field strength (N).
Weight changes from place to place but mass remains the same everywhere.

Moving Bodies

Concept Mapping

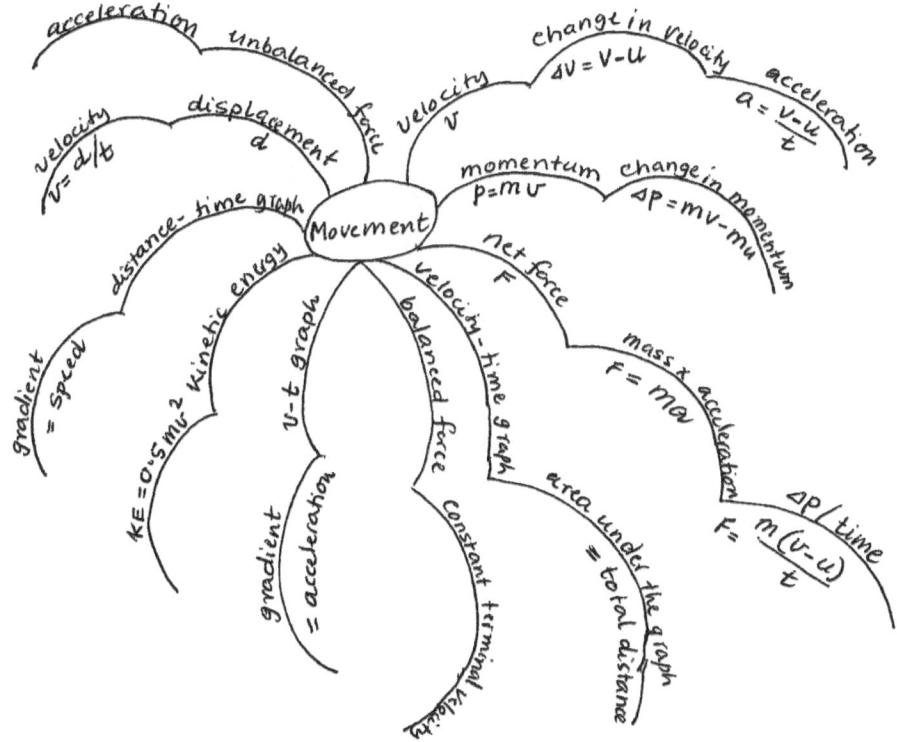

Speed = distance/time. It is a scalar quantity. It has size only. The unit of speed is m/s.

Velocity = distance/time. It is a vector quantity. It has size and direction. The unit of velocity is m/s.

Acceleration occurs when an object changes velocity. Speeding up is acceleration, and slowing down is deceleration. Acceleration or deceleration is calculated using the formula,

Acceleration or deceleration = change in velocity/time

$$|a| = \Delta v/t \; (\Delta v = \text{change in velocity}).$$

The unit of acceleration or deceleration is m/s^2.

Some Important Graphs

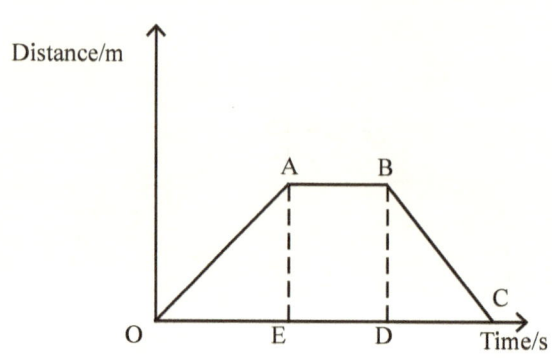

OA ⇒ constant velocity

AB ⇒ at rest, v = 0

BC ⇒ the body is going back at a constant velocity to the starting point.

Gradient of OA
= AE/OE = velocity.

Gradient of BC
= BD/DC = velocity.

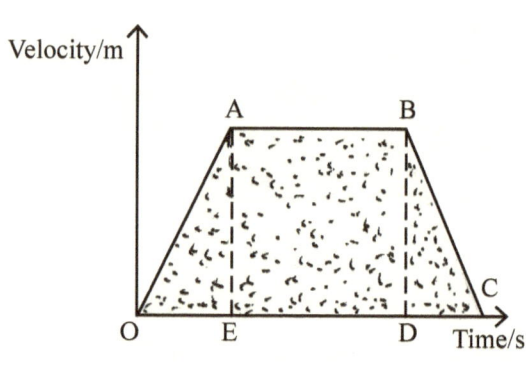

OA ⇒ constant acceleration

AB ⇒ constant velocity

BC ⇒ constant deceleration

Gradient of OA
⇒ AE/OE = acceleration

Gradient of BC
⇒ BD/DC = deceleration

Area under the graph = total distance travelled

Area under each section represents the distance travelled during that time.

Average velocity = total distance/total time

In this particular case, the area can be found from the area of the trapezium OABC.

Area of OABC = (AB + OC) × AE/2

Alternatively, the area can be obtained by calculating the area of the two triangles OAE and BDC and of the rectangle ABDE and adding them together.

$F = ma$. Therefore $F \propto a$, for a certain mass.

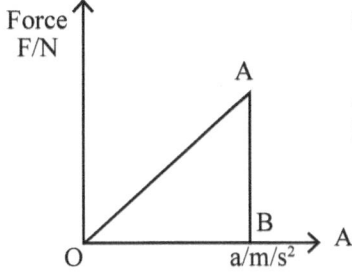

Gradient $\Rightarrow AB/OB =$ mass

Acceleration increases proportionately with increase of force for a fixed mass.

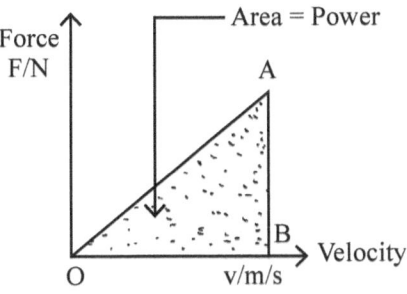

The greater is the force, the greater is the velocity for a fixed mass. The area under the F-v graph gives power.

Power = energy/time.

The unit of power is joule per second (J/s).

Force × velocity gives (N × m/s) = Nm/s = joule/second = J/s = power.

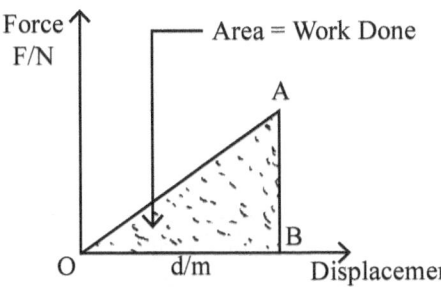

The displacement is directly proportional to the force applied on a fixed mass. The greater is the force, the greater is the displacement. The area under the F-d graph gives the amount of work done.

Force × distance gives Nm = joules (J) = work or energy.

While driving a car, the driver has to think and brake before the car comes to a stop when he/she sees any signs of hazard.

Thinking distance is the distance travelled by the car in the time interval when the hazard signal appears in the driver's view and when the brakes are applied. This distance depends on the speed of the car, the alertness of the driver, and visibility controlled by the weather conditions. Braking distance is the distance travelled by the car after the brakes have been applied until the car stops.

Stopping distance = thinking distance + braking distance.

Braking distance depends on

- the speed of the car
- the total mass of the car and the passengers
- the condition of the brakes
- the condition of the tyres
- the weather condition

It is clear that speed is the major factor in increasing the thinking and braking distances, thereby making the total stopping distance longer.

Whilst in motion, the car has kinetic energy $\frac{1}{2} mv^2$. Braking force opposes the motion of the car. The amount of work done by this braking force is $F \times d$, where d is the braking distance. Therefore, to stop the car,

$$1/2 \, mv^2 = F \times d.$$

Because of the term v^2, if the speed is doubled, the braking distance becomes four times for the maximum braking force.

Moving bodies possess momentum and kinetic energy. During collisions, total momentum before collision is equal to the total momentum after the collision. This is called the principle of conservation of momentum.

Momentum (p) = mass × velocity

p = mv. Momentum is a vector quantity (kg m/s)
Kinetic energy = ½ m v² (J).

Elastic collision ⇒ momentum is conserved, and KE is conserved.
Inelastic collision ⇒ momentum is conserved, but KE is not conserved.

Therefore, it is important to calculate momentum and KE before and after the collision to check whether the collision is elastic or inelastic.

Learning Zone

Relationship between

Distance (d), velocity (v), and time	d = v × t.
Force (F), mass (m), and acceleration	F = m × a.
Power (P), force (F), and velocity (v)	P = F × v.
Work (W), force (F), and distance (d)	W = F × d.
Energy (E), power (P), and time (t)	E = P × t.

Bodies in motion
with a constant acceleration,

where,

v = u + at

s = ut + ½ at²

v² = u² + 2as

a = (v − u)/t

v = final velocity in m/s
u = initial velocity in m/s
t = time in seconds
s = distance in m
a = acceleration in m/s².

From the acceleration equation, the force equation can be changed to other forms containing momentum.

F = ma

F = m (v − u)/t

F = (mv − mu)/t

Ft = mv − mu

Here, mv = final momentum (kg m/s)
mu = initial momentum (kg m/s)
mv-mu = change in momentum (kg m/s)
Ft = inpulse (Ns)

Matter Matters

Concept Mapping

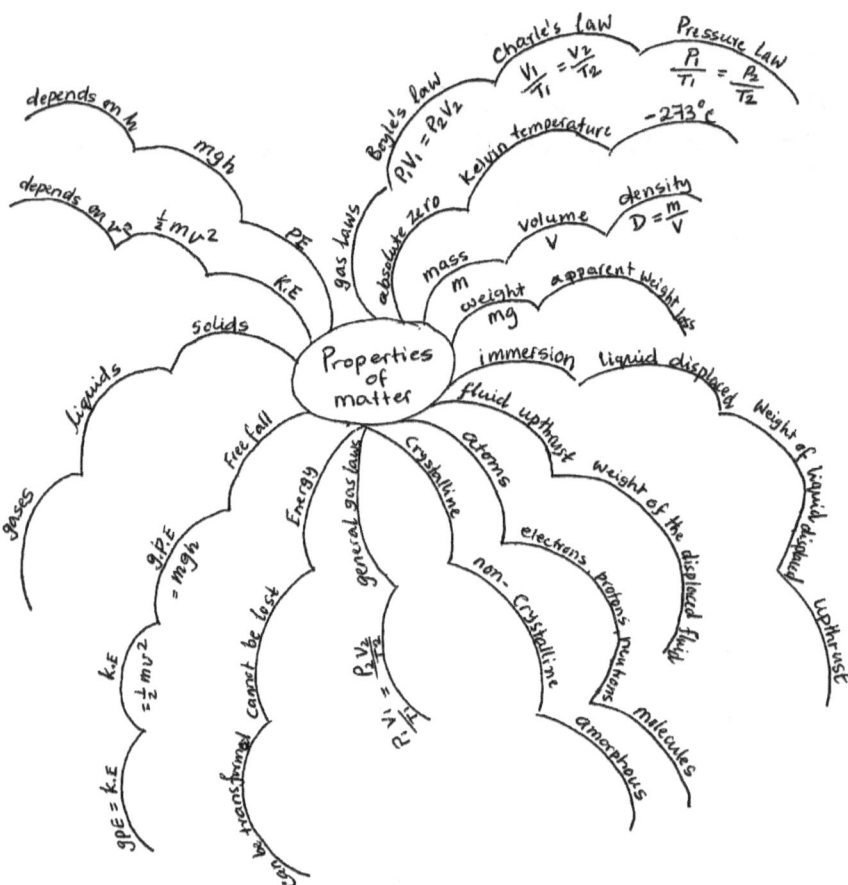

Density

Instead of saying an object is lighter or heavier, we should say whether the density of a substance is less or more compared with another substance.

Density = mass (kg)/volume (cm³), or density = mass (g)/volume (cm³)

The unit of density is kg/m^3 or g/cm^3.

Hot air has lower density than cold air. Hence hot air rises upwards and cold air falls downwards.

Similarly, hot water has lower density than cold water. Therefore, hot water rises to the top and cold water falls. In both the cases, convection current is set up until an equilibrium temperature is reached.

Apparent Weight Loss

When an object is fully immersed in water, it displaces its own volume of water.

The upthrust on the object is equal to the weight of the displaced water.

Upthrust causes the apparent weight loss.

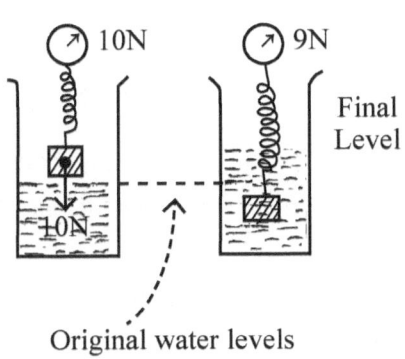

Original water levels

g is taken to be 10 N/kg.
Density of water = 1 g/cm^3.

Volume of water displaced = 100 cm3.
Mass of water displaced= 100 g.
Weight of water displaced = 1 N.
Upthrust = 1 N.

So the newton meter on the right shows (10-1) = 9 N.

Energy

Capacity to do some work is called energy.
Force × distance moved in the direction of the force (d) = F × d = work (W)
    Work ≡ energy
Energy transfer per second is power (P)

Therefore $P = E/t$ or $P = W/t$.
Energy is in joules (J). 1000 J = 1 kJ.
Power is written in J/s or watts. 1000 watts = 1 kW

Gravitational potential energy (gpe) is due to the height of the object from the ground level.

gpe = mass (kg) × gravitational field strength (g) × height (h) = mgh
    (J).
Kinetic energy is due to the motion of a body.
KE = ½ × mass (kg) × v^2 (m^2/s^2) = ½ mv^2 (J)

According to the principle of conservation of energy, energy cannot be destroyed, but it changes from one form to another.

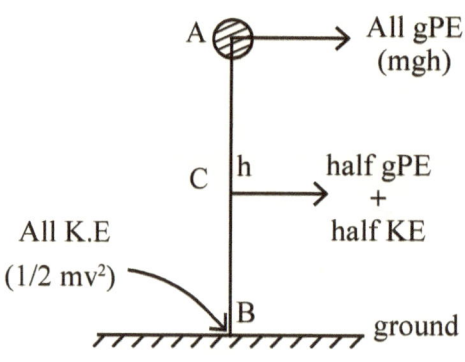

At A, an object of mass m possesses all potential energy.

At the midpoint C, it has half gpe and half KE.

At B, the object has only KE.

All the gpe at A has now been converted into KE at B.

If no energy is lost during this fall,
    KE = gpe
    ⇒ ½ m v^2 = mgh
    ⇒ v^2 = 2gh
    ⇒ v = √(2 gh).

This gives the velocity of the object just before it touches the ground. Remember, on touching the ground, heat and sound energy will be produced.

Since g is the gravitational field strength, which is a constant, the velocity depends on height only.

Pressure

Pressure is the force applied on 1m².

    Pressure = force/area (N/m²)

For a liquid of depth h,
    P = depth × density × gravitational field strength
    P = h d g (Pa).

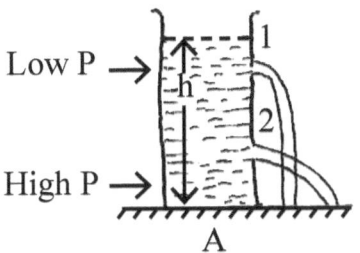 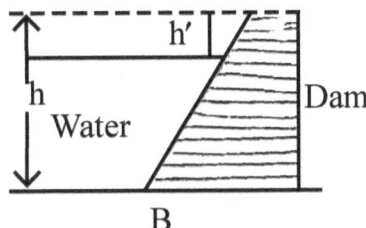

In diagram A, water comes out to a short distance from hole 1, but from hole 2, water comes out to a longer distance due to their respective heights from the top level.

In diagram B, the dam has a wider base because the pressure is high at the bottom due to greater depth. The top of the dam is thinner as the water pressure is low at the top.

Pressure can be measured using a pressure gauge. Atmospheric pressure, which controls our weather conditions, is measured with a barometer.

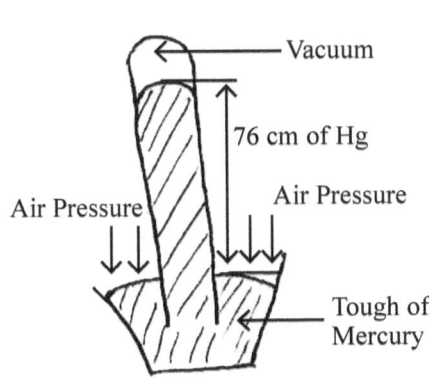

One atmospheric pressure is equal to a vertical height of 76 cm of mercury (Hg).

The atmospheric air pushes the mercury in the trough, which, in turn, holds the length of the mercury column in the glass tube.

High pressure ⇒ good weather
Low pressure ⇒ bad weather

The vertical height of the mercury column gives the atmospheric pressure.

The atmospheric pressure does not depend on the diameter of the glass tube.

In an aneroid barometer, there is no liquid.
In a liquid, pressure acts equally in all directions.

Pressure Transmission

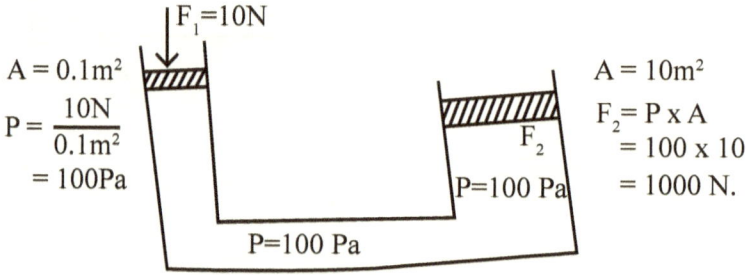

Area ratio $\Rightarrow$ 0.1:10   Force ratio $\Rightarrow$ 1:100
         $\Rightarrow$ 1:100 $\Rightarrow$ 10:1000.

By applying a small force in a smaller area on the left-hand side of the system, a much bigger force is produced on the right-hand side of the system, which then allows a bigger load to be lifted up on the bigger area side.

Matter Exists in Three States

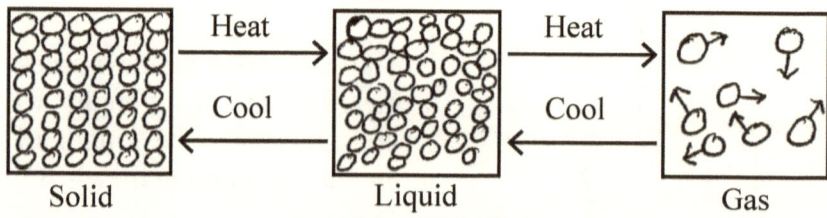

Matter can be thought to be made up of particles for our understanding.

Solids

Particles are closely packed.
Particles stay in one position, but can vibrate.
Particle vibration becomes more at higher temperatures.
A large number of particles are present in a small volume.
Strong force exists between the particles.

Liquids

The force between the particles is weak.
Particles can move around.
Particles are in a constant motion in all directions.
Particles are less closely packed.
A liquid had less density than a solid.

Gases

Particles are far apart.
There is no force of attraction between particles.
Particles are moving and colliding with each other and the wall of
    the container at all times.
Density of a gas is less than those of liquids and solids.
Gases can be compressed easily.

Cooling graph of a Substance

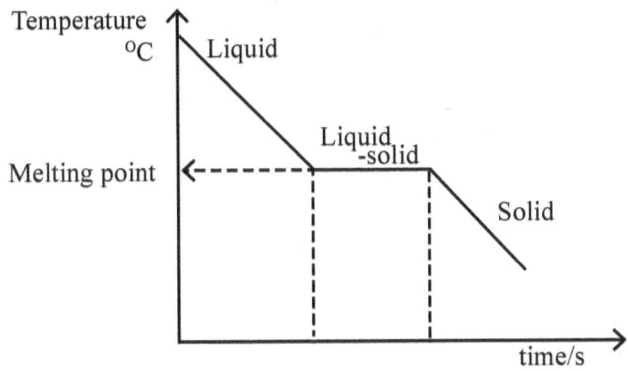

The cooling graph shows that during change of state, there is no change in temperature. The extra thermal energy needed to change the state at a constant temperature is called latent heat.

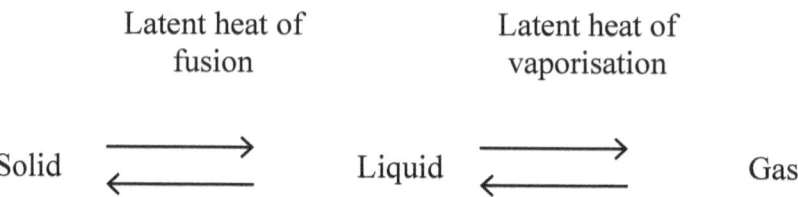

All matter consists of molecules. At any temperature, the molecules have certain energy and they vibrate. In a solid, they vibrate in their own position in the crystalline state, but they do not move. In a liquid, they vibrate and move from their positions but not very far apart. In a gas, however, the molecules are moving around randomly at all possible speeds and in all possible directions.

All matter expands on heating. Therefore, when heated, the volume becomes more for a given mass. Since density = mass ÷ volume, the density becomes less when the temperature is high because the volume becomes bigger.

A gas is always kept in a container in which it occupies the whole volume. By virtue of their kinetic energy and random motion, the molecules hit the walls of the container. This is a force which causes the pressure of the gas. For a given mass of a gas in a given container, the pressure of the gas becomes more when we heat the gas. This is because at a higher temperature, the molecules have more kinetic energy to hit the walls of the container with more force on a given area.

Three quantities define the properties of a gas:
Pressure (P), volume (V) and temperature (T).

There are three gas laws:

Boyle's law $\Rightarrow$ at a constant temperature (isothermal),
$$P \propto 1/V \Rightarrow PV = \text{constant}. \; P_1 V_1 = P_2 V_2.$$
Charles's law $\Rightarrow$ at a constant pressure (isobaric),

$$V \propto T \Rightarrow V/T = \text{constant. } V_1/T_1 = V_2/T_2.$$

Pressure law $\Rightarrow$ at a constant volume (isovolumetic)

$$P \propto T \Rightarrow P/T = \text{constant. } P_1/T_1 = P_2/T_2.$$

General gas law $\Rightarrow$ PV/T = constant. $P_1V_1/T_1 = P_2V_2/T_2$.

In all the above equations, T represents Kelvin temperature.
$^\circ$C + 273 = Kelvin temperature (K).

Boyle's Law (Temperature Constant)

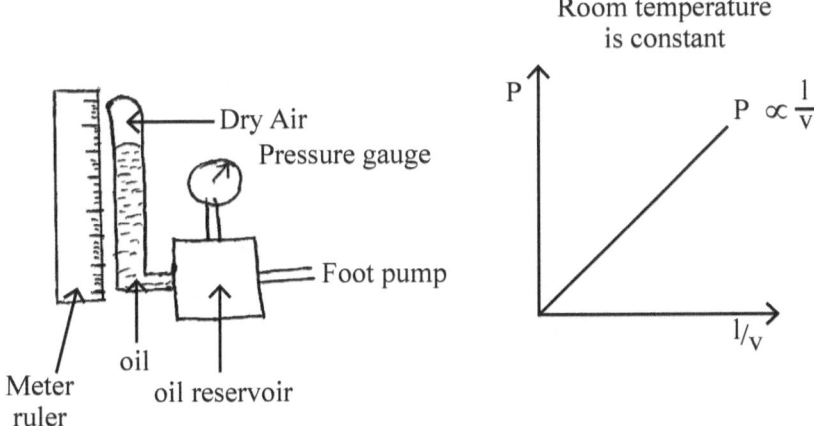

The length of the dry air is proportional to the volume. By applying the foot pump, the pressure is varied, and the corresponding length is measured by the ruler.

Charles's Law (Pressure Constant)

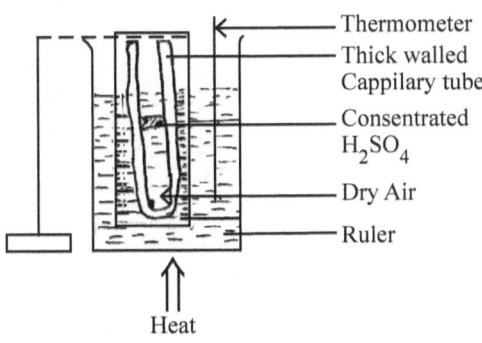

The atmospheric pressure over the concentrated H2SO4 pellet is constant.

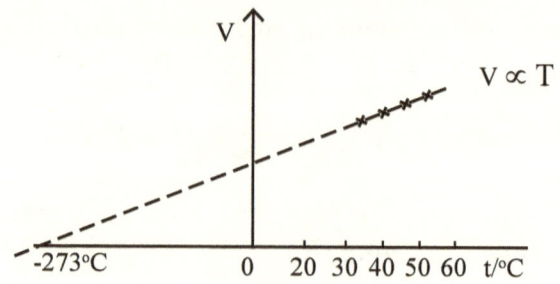

Pressure Law (Volume Constant)

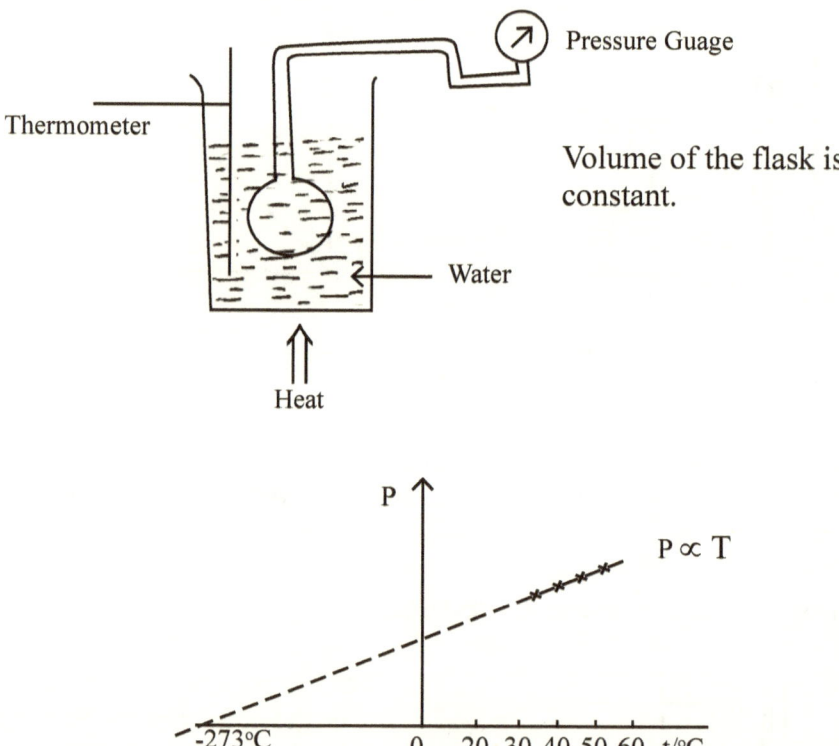

Volume of the flask is constant.

Points to Remember

What is kept constant and what are the variables?

What are the variables to plot a graph to get a straight line? This can confirm direct proportionality of the two quantities.

Approximate values of the temperatures and show $-273°C$ [OK] by extrapolating the straight line.

Brownian Motion

Smoke cell experiment $\Rightarrow$ some smoke particles are introduced into the cell and covered with the microscope slide. Through the microscope, we see the movement of the bright specks, which are smoke particles bombarded by the air particles. It represents the movement of gas particles, which is random and in all possible directions.

Diffusion

Liquid bromine is poured into the bottom of a gas jar, and the glass jar is covered with a lid. Leave the jar for about half an hour. The jar will be filled with brown bromine gas. The bromine gas particles move through the spaces available between the air particles, causing diffusion.

Learning Zone

Density = mass/volume (kg/m^3 or g/cm^3)

Work done = force $\times$ distance = F $\times$ d (J)

Gravitational potential energy = mass $\times$ height $\times$ gravitational field strength = mgh (J)

Kinetic energy = $\frac{1}{2} \times$ mass $\times$ (velocity)^2 = $\frac{1}{2}$ mv^2 (J)

Equating KE = g PE gives v = $\sqrt{(2gh)}$

Liquid pressure = depth $\times$ density $\times$ gravitational field strength = hdg (Pa)

Boyle's law (temperature constant) $\Rightarrow P_1V_1 = P_2V_2$

Charles's law (pressure constant) $\Rightarrow V_1/T_1 = V_2/T_2$

General gas law $\Rightarrow P_1V_1/T_1 = P_2V_2/T_2$

Pressure Law (volume constant) $\Rightarrow P_1/T_1 = P_2/T_2$

Energy and Energy Transfer

Concept Mapping

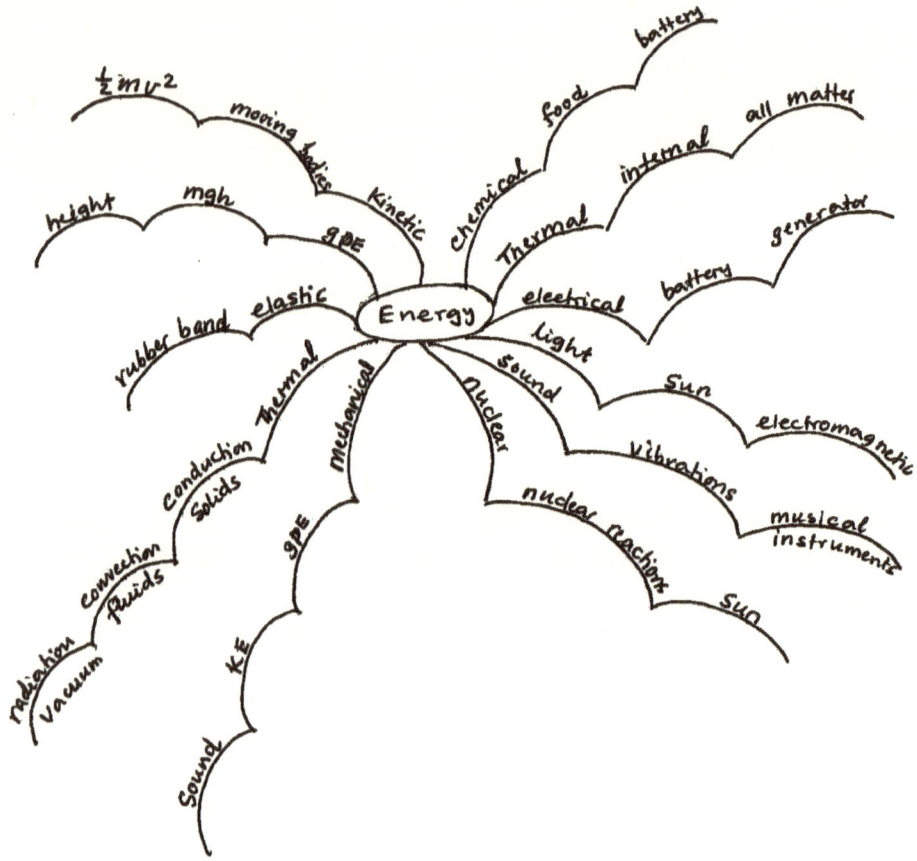

Conduction in solids is due to the vibration of the particles and movement of the free electrons.

Convection in liquids and gases is due to the convection current set up by the hot and cold particles. Hot gas particles and hot liquid particles rise, and cold gas particles and cold liquid particles fall.

Radiation is an electromagnetic wave which can travel through vacuum. Sun's radiation travels to the earth through the space.

Energy Transformation

Light bulb   ⇒ electrical energy   → light and heat
Microphone   ⇒ sound energy   → electrical
Loud speaker ⇒ electrical energy   → sound
Dynamo   ⇒ kinetic energy   → electrical
A falling body ⇒ gpe   → kinetic

Work done is an energy transfer process.

A machine does work and produces output energy. All machines have three main parts – input, processing, and output.

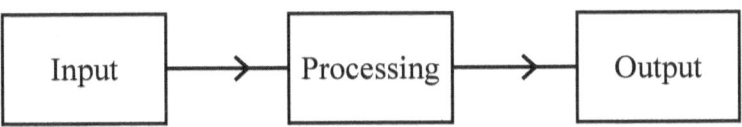

The output power or energy is always lower than the input energy because of the energy lost in other forms during processing.

$\square$ efficiency $\square\dfrac{\text{output energy or power}}{\text{input energy or power}}$ $\square\ \square\square\square$

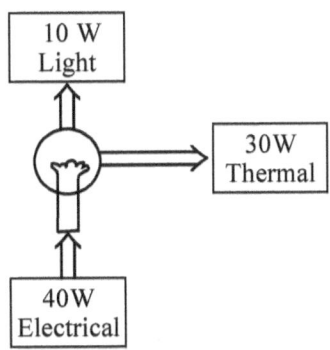

A $\square$W light bulb gives only 1$\square$W of useful light energy, and 3$\square$W is lost as thermal energy.

$\square$ efficiency $\square$ $\square\square\square\square\square\square\square\square$ $\square\square\square\square$ $\square$ $\square\square\square$

To find the power developed by a person, you have to find the amount of work done in a given time. The work done is found from the formula$\square$

Work done $\square$ weight of the person $\overline{\text{m}}$g$\square\square$ vertical height.

If we find the vertical height of a staircase and the time taken by a person in seconds to climb up the staircase, we can work out the power developed by the person from:

Power = work done/time in seconds (J/s) or watts.

Section 2

Waves and Vibrations (I)

Concept Mapping

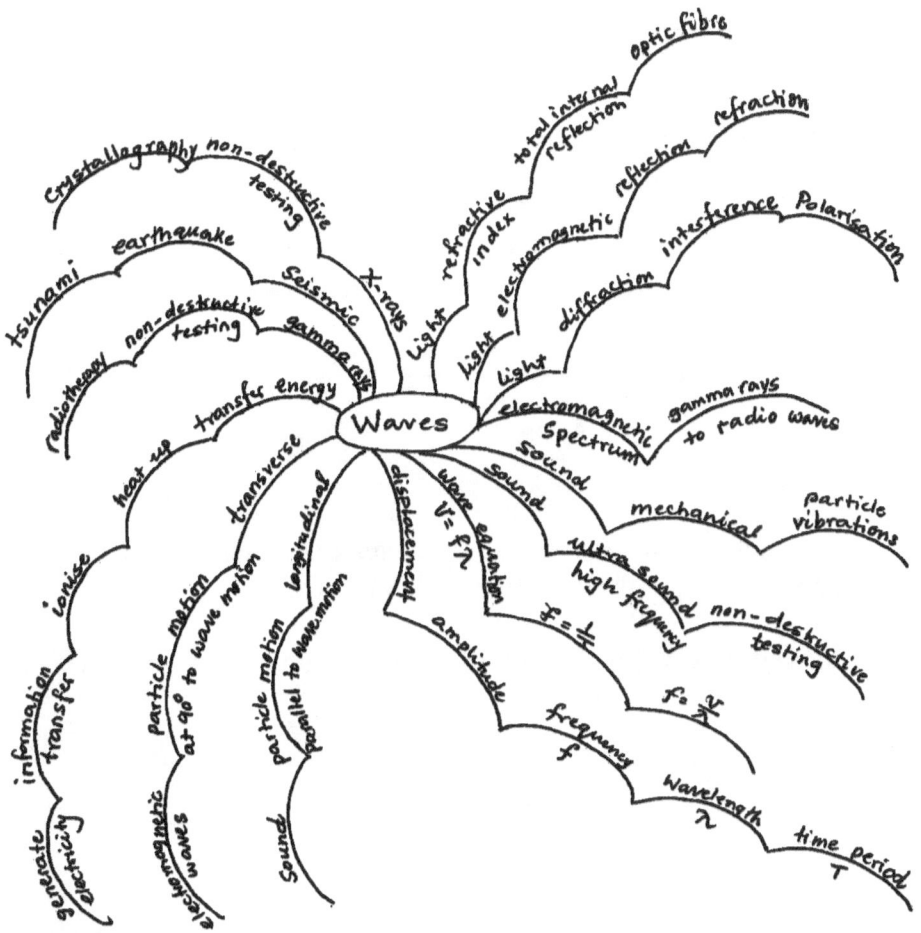

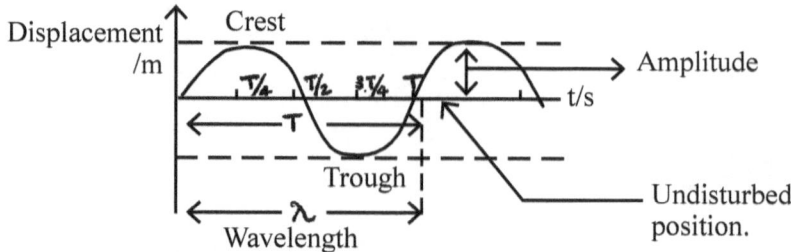

Time period is the time taken in seconds for one complete vibration. This is written as T.

Frequency (HZ) is the number of complete vibrations in one second. So, f = 1/T.

Wave equation $\Rightarrow$ velocity (m/s) = frequency (Hz) × wavelength (m)
$$\Rightarrow v = f\lambda$$

Transverse Waves: The particles of the medium vibrate perpendicularly to the direction of wave motion.

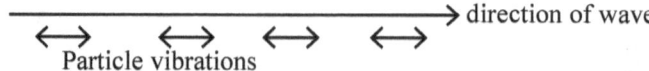

Particle vibrations

All electromagnetic waves are transverse. Our light is a transverse electromagnetic wave. Transverse waves can be polarised. That means they can be made to vibrate in one particular plane.

Longitudinal Waves: The particles of the medium vibrate parallel to the direction of wave motion.

Sound is a mechanical wave in which the particles vibrate in the same direction as that of wave motion. So it is a longitudinal wave. Sound waves cannot be polarised. Sound waves cannot travel through vacuum as there are no particles to produce vibrations.

The speed of a sound wave depends on the density of the material through which it moves.

Density of air = 0.001 g/cm³   Speed of sound = 330 m/s
Density of water = 1.0 g/cm³   Speed of sound = 1400 m/s
Density of iron = 7.9 g/cm³   Speed of sound = 5000 m/s

Sound waves reflect from hard surfaces such as walls in a room. They are also absorbed by materials like carpets, cushions, settees, and curtains. When we speak in an empty room, the echoes produced result in reverberation. But in a well-furnished room, reverberation does not occur due to sound absorption by the furnishing materials.

When the amplitude is greater, the sound is louder.
When the frequency is higher, the sound has a high pitch.
When the frequency is lower, the sound has a low pitch.

We can hear sounds of frequency from 20 Hz upto about 20000 Hz or 20 kHz.

Sound of frequency higher than 20 kHz is called ultrasound.

Ultrasound has many applications.

It is used in industrial cleaning because of its high frequency vibrations, e.g. cleaning teeth, and medical tools. It is used in breaking kidney stones and gallstones without the necessity of any major surgery.

Ultrasound is used for scanning a fetus in mother's womb. Various reflections obtained from different areas of the fetus are processed by a computer, and a video image is produced.

Bats send high-pitched sounds, which then reflect from their prey and are received by their brain. They can then follow the path of the reflected signal to catch their prey.

Ultrasound is used to find the depth and features of the seabed by similar echo reflection technique.

Reflection of Light

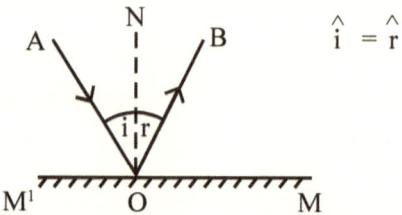

ON   ⇒ Normal at 90° to the mirror MM′.
i   ⇒ Angle of incidence.
r   ⇒ Angle of reflection.
AO   ⇒ Incident ray.
OB   ⇒ Reflected ray.

Refraction of Light

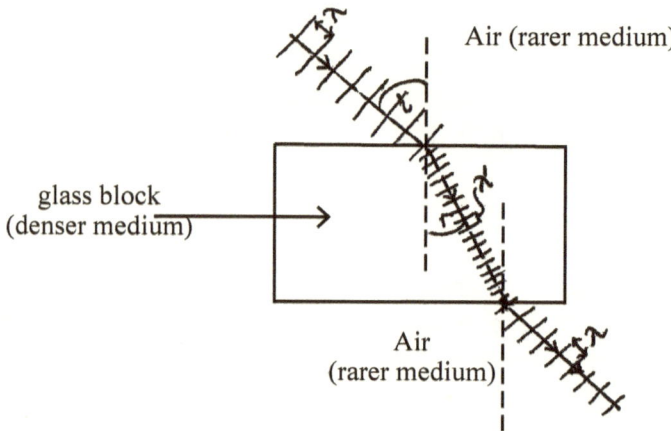

Light is travelling from rarer medium air to a denser medium glass and out again into air. The direction of the ray changes at the interfaces.

Velocity of light in air $= 3 \times 10^8$ m/s.
Velocity of light in glass $= 2 \times 10^8$ m/s.
Velocity in glass is less than velocity in air.
Wavelength in glass (λ') is less than wavelength in air (λ)
Frequency of light does not change.

Critical Angle and Total Internal Reflection

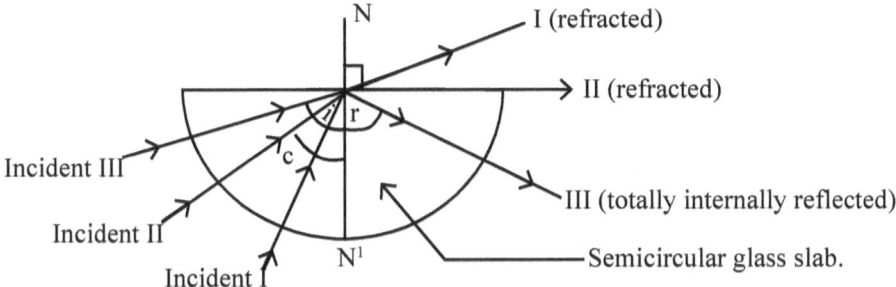

Ray I $\Rightarrow$ Ordinary refraction.

Ray II $\Rightarrow$ The refracted ray is leaving by the surface of the slab making an angle of 90° with the normal.

$\hat{C} \Rightarrow$ Critical angle between incident ray II and the normal, inside the block.

Ray III $\Rightarrow$ The angle between incident ray III and the normal is bigger than critical angle C. The ray is then totally internally reflected inside the glass following the laws of reflection. This is called total internal reflection.

Fibre-optic cables for transmitting audio and video signals work on the principle of total internal reflection.

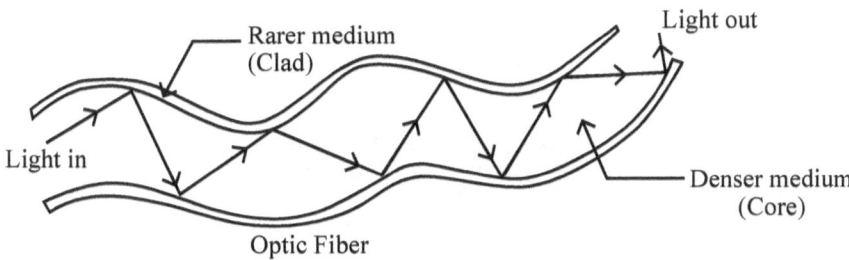

Interference: Two exactly alike waves (having same v, f, and λ) superimpose to produce maximum and minimum, resulting in interference.

Crests falling on crests produce maxima (constructive interference).
Troughs falling on troughs produce maxima (constructive interference).

Crests falling on troughs or vice versa produce minima (destructive interference).

Interference of sound waves can be demonstrated as in the following diagram.

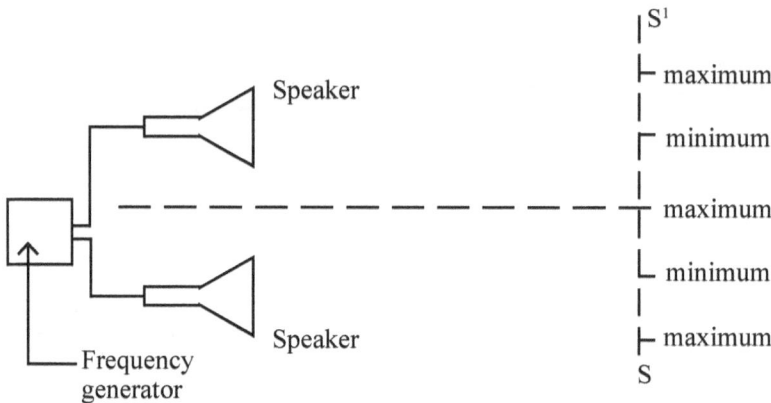

Waves coming out of the speakers with same f and λ diffract and superimpose and produce maximum and minimum sound a few metres in front as a result of interference. A person walking slowly along the line SS′ will hear maximum and minimum sound.

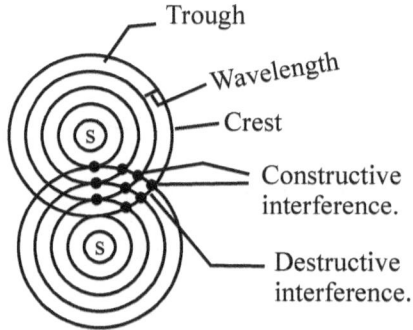

These two sources S and S′ having the same v, f, and λ are producing spherical waves. They superimpose and produce maximum and minimum as a result of interference. These two sources may be two spherical dippers connected to an eccentric motor driven by a frequency generator in a ripple tank apparatus, using water to demonstrate interference.

Interference of Light

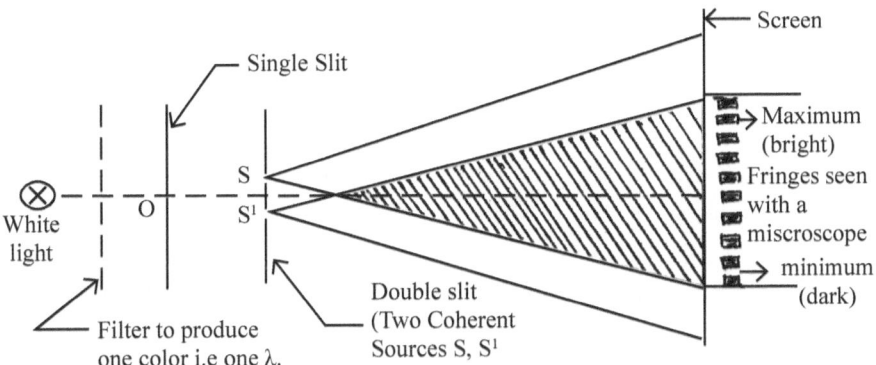

Two cones of light coming out from S and S' superimpose in the shaded area and dark and bright fringes can be seen on a screen or in the eyepiece of a microscope. The two sources S and S' are produced from one single source at O. Therefore, the two sources are coherent because they have the same frequency and wavelength. The filter produces one colour, eliminating the other colours from the white light. So, light comes out from the single slit with one frequency and one wavelength.

Waves from the two coherent sources arrive at different points on the screen in the hatched area, travelling different distances. The difference between the distances covered is called the path difference measured in terms of wavelength.

If the path difference = $O, \lambda, 2\lambda, 4\lambda$, etc., we get constructive interference with a maximum.
If the path difference = $\lambda/2$, $3\lambda/2$, $5\lambda/2$, etc., we get destructive interference with a minimum

Diffraction

When there is a bending of a wave around a corner, or when the wave passes through a small gap, diffraction occurs. The gap through which the wave passes must be smaller than the wavelength of the wave to produce diffraction.

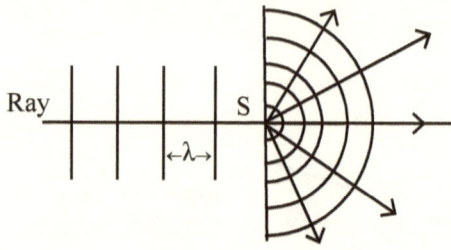

S is a small gap compared to the wavelength λ. So diffraction occurs.

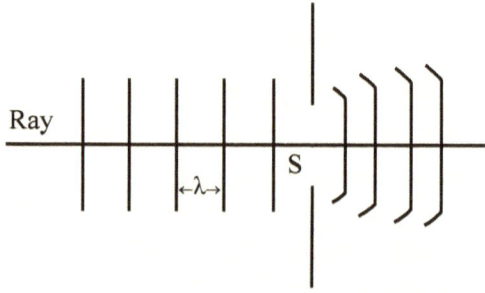

In this case S is a bigger gap compared to the wavelength λ. There is hardly any diffraction taking place.

Electromagnetic Spectrum

Concept Mapping

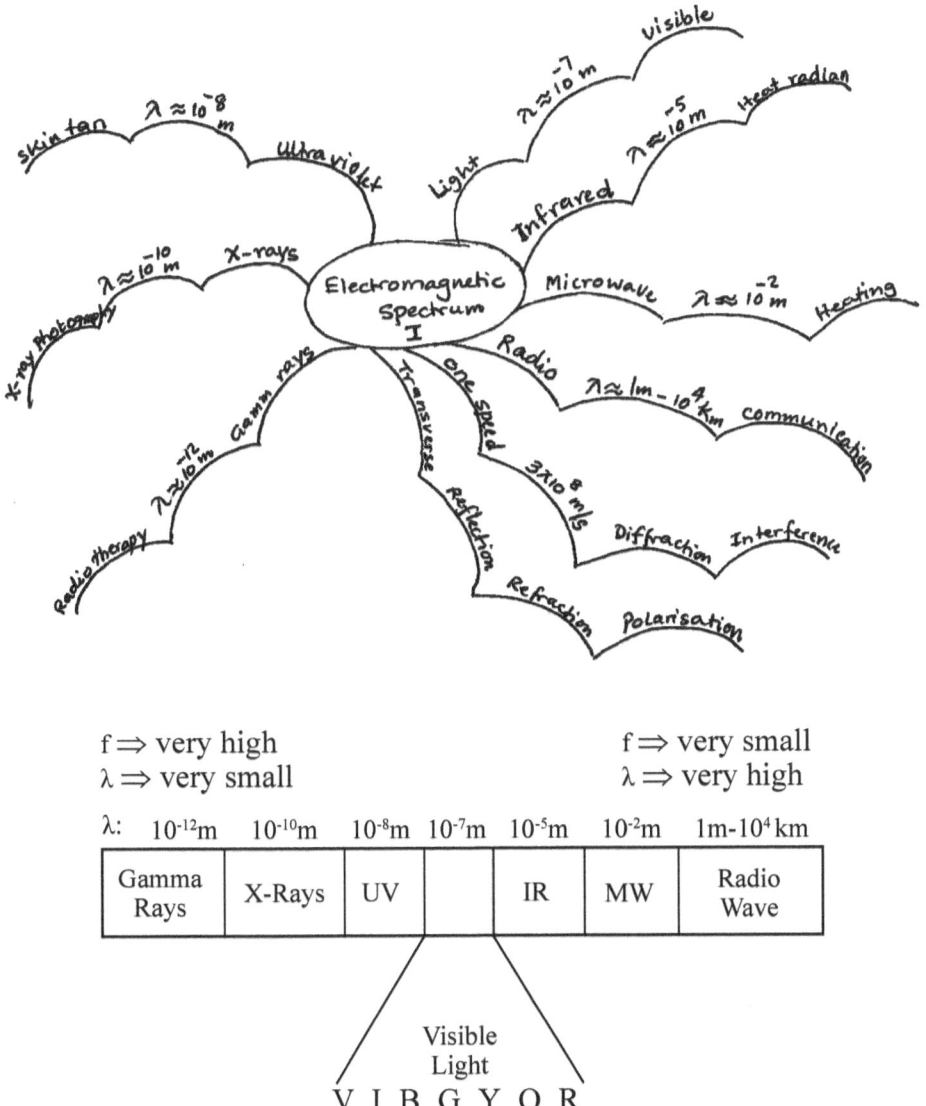

f ⟹ very high   f ⟹ very small
λ ⟹ very small   λ ⟹ very high

λ:	10^{-12}m	10^{-10}m	10^{-8}m	10^{-7}m	10^{-5}m	10^{-2}m	1m-10^4 km
Gamma Rays	X-Rays	UV		IR	MW	Radio Wave

Visible
Light
V I B G Y O R

Seven colours – violet, indigo, blue, green, yellow, orange, and red – are present in our visible light, from low wavelength to high wavelength, in that order. These colours can be remembered in that order using the sentence:

<u>V</u>ery <u>I</u>ntrepid <u>B</u>anker <u>G</u>ets <u>Y</u>elling <u>O</u>ver <u>R</u>eport.

Properties of Electromagnetic Waves

Concept Mapping

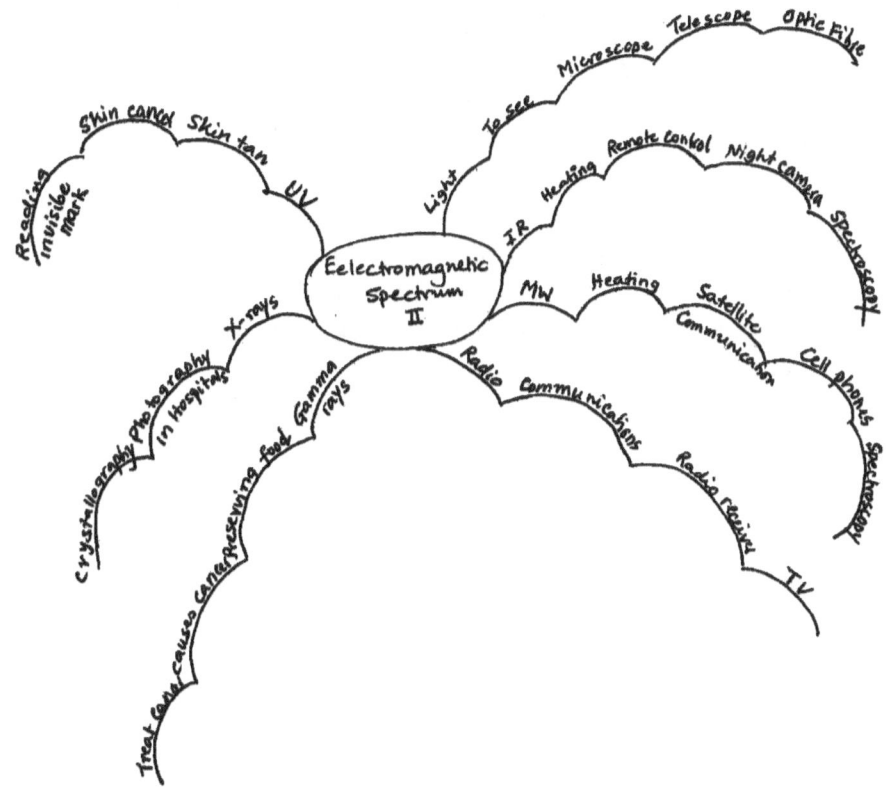

Sources of Electromagnetic Spectrum

Radio wave	$\Rightarrow$	Radio transmitter
Microwave	$\Rightarrow$	Klystron
Infrared	$\Rightarrow$	Hot bodies
Light	$\Rightarrow$	Sun
Ultraviolet	$\Rightarrow$	Sun
X-rays	$\Rightarrow$	X-ray tube
γ-rays	$\Rightarrow$	Radioactive isotope

Dispersion

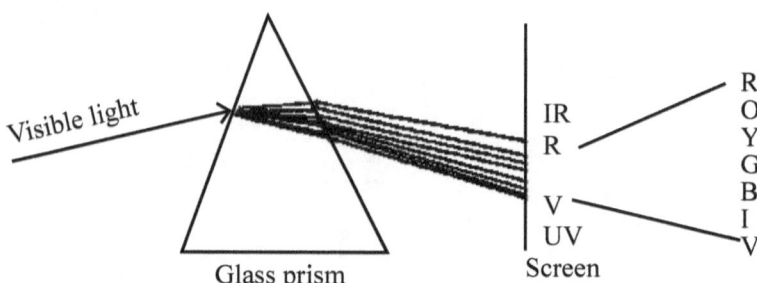

A glass prism separates the seven different colours on to the screen. This is called dispersion. Different wavelengths are deviated through different angles due to refraction producing a spectrum on the screen. Red colour is least deviated and violet the most.

The Earth and Seismic Waves

Concept Mapping

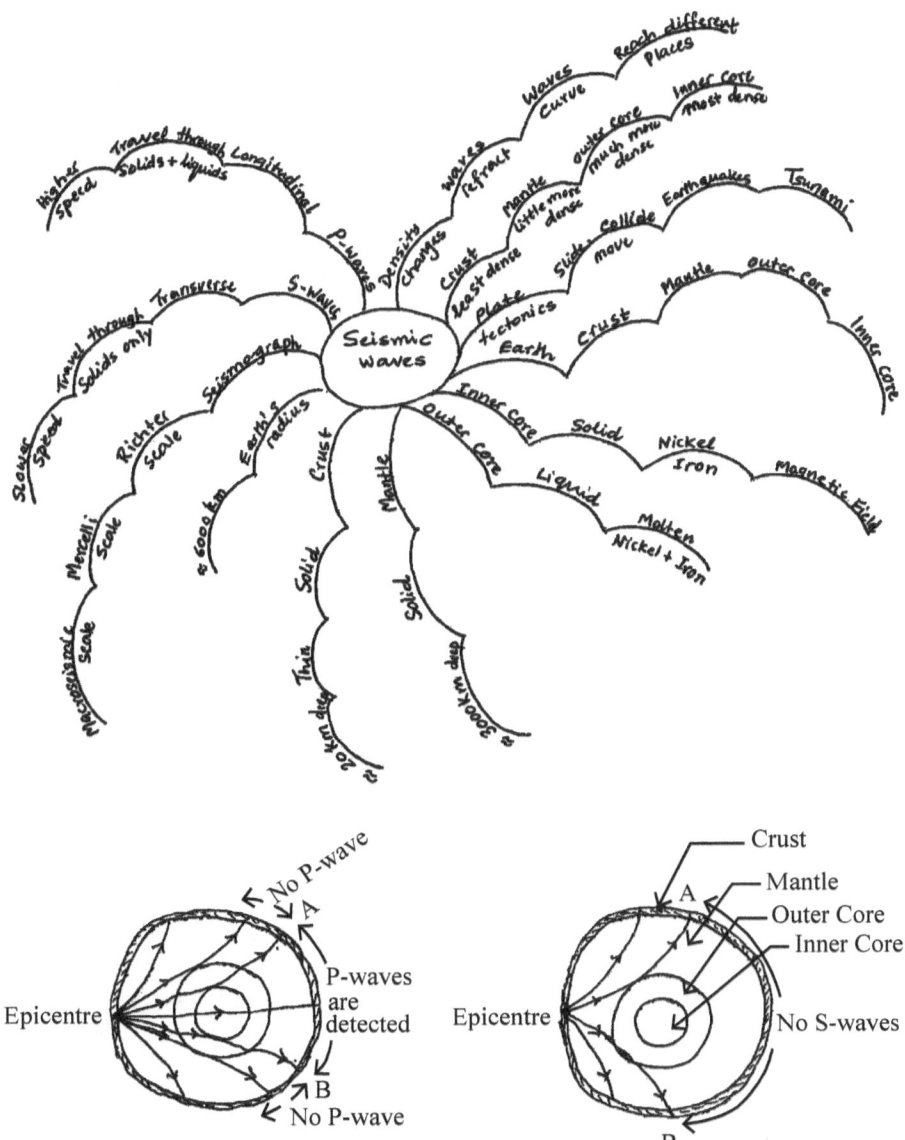

Longitudinal P-waves go through solids and liquids to reach the regions between A and B where they can be detected. P-waves travel faster than S-waves.

Transverse S-Waves travel through solids only. So they cannot reach the region between the points A and B. S-waves travel slower than P-waves.

Whilst P-waves produce up and down vibrations to any structure, S-waves make the sideways movements of the structures. The intensity of an earthquake is measured by a seismograph using the Richter scale. Modified Mercalli scale and the Macroseismic scale have been used in recent years.

Section 3

Movement of Electric Charge

Concept Mapping

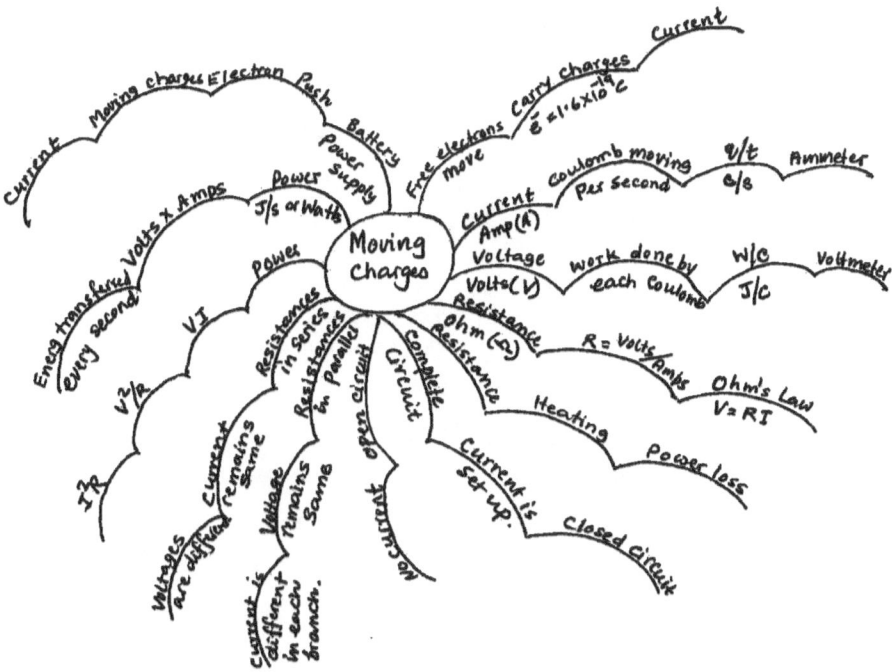

Moving charges produce electric current. Electrons are the charge carriers. Each electron has a charge of 1.6×10^{-19}C. In a conductor, there are a huge number of free electrons. In a battery, stored chemical energy gets converted into electrical energy on completion of the electric circuit. This provides the electrical push for the sea of free electrons to move in the connecting wires, and a current is set up.

Electrons have negative charges. Remember, both positive and negative charges can produce current when they move.

Current (I) is measured in amps (A) by an ammeter which is connected in series with the circuit.

Voltage (V) is measured in volts (V) by a voltmeter which is connected in parallel to the component, e.g. a bulb, a resistor.

Resistance (R) is measured in ohms (Ω) by an ohmmeter or by calculation from the equation: $V = RI$ (Ohm's Law).

Series Circuit

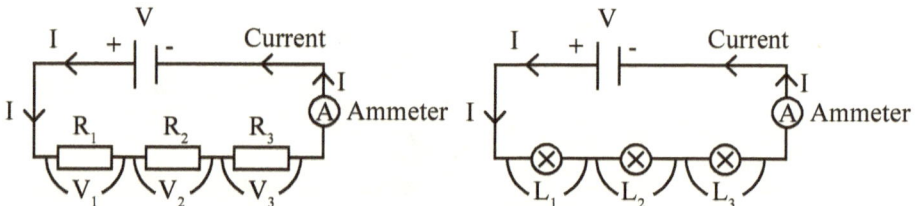

Resistors connected in series   Light bulbs connected in series. They
                                       have resistance in the filament

Total resistance $R_T = R_1 + R_2 + R_3$. If one bulb blows, the other bulbs do not work since the circuit is broken. The current is same at all points in the circuit. $V = V_1 + V_2 + V_3$. Voltage across each resistor is different.

Parallel Circuit

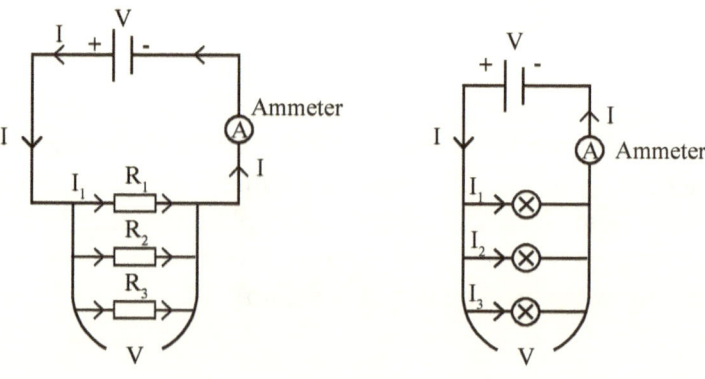

Voltage across each resistor or each bulb is same (V), but current is different through each resistor or each bulb. The main current is measured by the ammeter (A). If all the resistors or the bulbs are same, then each branch will receive one-third of the main current. The voltage across each resistor or each bulb remains same (V), which is equal to the supply voltage V. If the resistors or the bulbs are different, bigger current passes through smaller resistance and vice versa. Each current can be calculated from the formula $I = V/R$.

Total resistance is given by the equation $1/R_T = 1/R_1 + 1/R_2 + 1/R_3$.
For two parallel resistors $1/R_T = 1/R_1 + 1/R_2 \Rightarrow R_T = R_1 R_2 / (R_1 + R_2)$.

If one bulb blows, the others will still work, and they will have the same brightness as before. Household lightings are connected in parallel.
A longer wire has greater resistance than a shorter wire.
A wire with a larger diameter has smaller resistance than a wire with a smaller diameter.

$R \propto l \Rightarrow$ Resistance doubles when the length is doubled.
$R \propto 1/A \Rightarrow$ When the cross-sectional area doubles, resistance is halved.
$R \propto 1/\pi r^2 \Rightarrow$ Resistance becomes a quarter when the radius is doubled.

Fuses are used to control the current in an appliance. The current rating in a fuse is calculated from $I = P/V$. We use fuses of one value higher than the calculated value. Fuse is a safety mechanism for an appliance.

Earthing is a safety device for the person handling the appliance and also for the appliance itself. When a surge of electricity develops, it passes to the earth through the earth wire, and at the same time, the fuse blows. Normally, the earth wire is connected to the water pipe under the sink in the kitchen. The casing of an appliance is connected to the earth wire.
Live$\Rightarrow$brown, neutral$\Rightarrow$blue, earth$\Rightarrow$yellow and green. The fuse is always connected in series with the live wire.

Fundamentals of Electricity

Chemical energy in the battery is converted into electrical energy in the connecting wires once the circuit is complete. The free electrons are thus moved in the circuit. The electrons move from the negative to the positive terminal of the battery. Electrons carry the charges which flow at a constant speed throughout the circuit.

Amount of charge (coulombs C) flowing per second is called current.

So, current I = charge/time = $C/t \Rightarrow C = It$
Current is measured in amps (A) or coulomb/second. $13A = 13C/s$.

When the charges flow through the circuit, they transfer energy to the components, e.g. a bulb, a resistor in the circuit. The amount of energy transferred in joules by each coulomb of charge is called a volt.

$1V = 1J/C \Rightarrow 1J$ of energy transfer by each coulomb.
$12V = 12J/C \Rightarrow 12J$ of energy transfer by each coulomb.
Power = voltage (J/C) × current (C/S) $\Rightarrow$ J/s or watt (W)
$1kW = 1000$ watts
$1kWh = 1kW$ used for an hour = $1kW \times 1h$.
Electrical energy = power (J/s) × time (seconds) $\Rightarrow$ J

$V = E/Q$	$E \Rightarrow VIt$	$P \Rightarrow VI$	
$E = VQ$	$E \Rightarrow I^2Rt$	$P \Rightarrow I^2R$	$V = RI$
	$E \Rightarrow (V^2/R)\,t$	$P \Rightarrow V^2/R$	

Ohm's Law

Voltage across a fixed resistor is directly proportional to the current passing through it when the temperature is constant.

$$V \propto I \Rightarrow V = RI$$

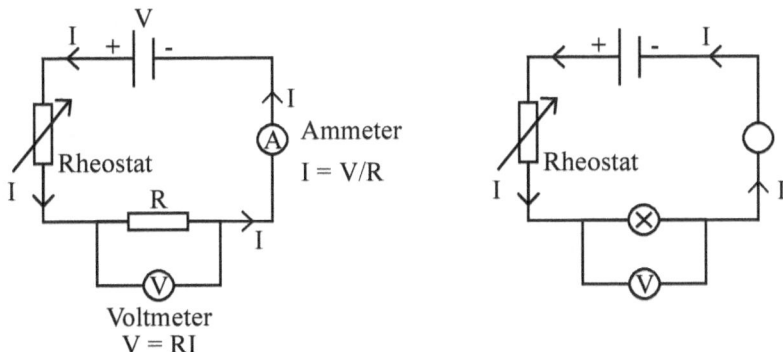

By changing the supply voltage V or by using a rheostat, the current can be changed in the main circuit.

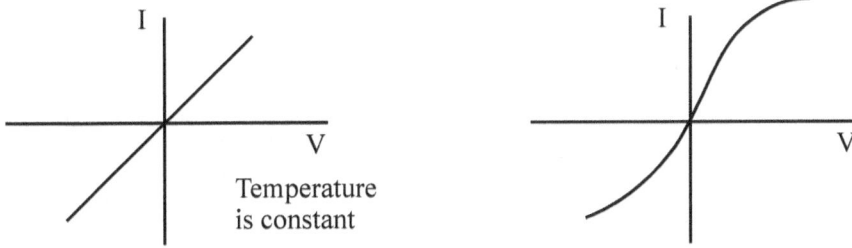

At a constant temperature, resistance of a conductor does not change. However, if the temperature of a conductor is changed, it causes its resistance to change. The higher the temperature, the higher is the resistance of a conductor. I–V graph for a filament bulb shows the effect of temperature on the resistance of the filament.

As the current is increased, the filament gets hotter and the atoms in the conductor vibrate more, opposing the flow of electrons. Hence the resistance increases.

For a thermistor, resistance decreases as the temperature increases because it is made up of a semi-conducting material. A thermistor is a temperature-dependent resistor.

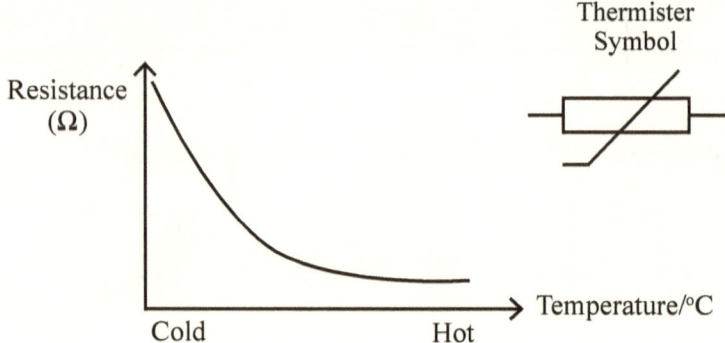

For an LDR (light dependent resistor), resistance decreases with increasing light intensity.

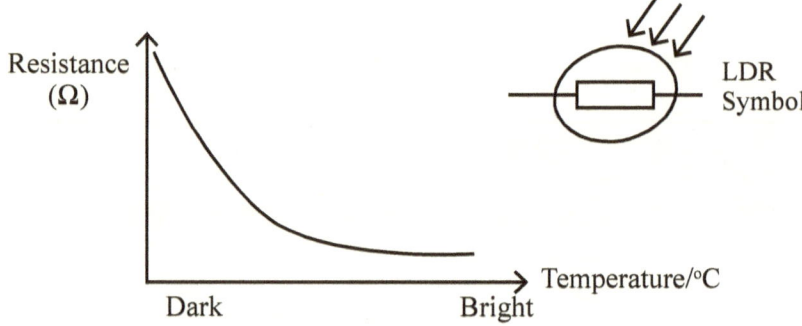

A diode is a semiconductor component which allows current to flow from the positive side to the negative only.

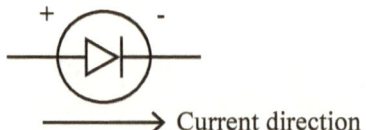

Current cannot flow in the reverse direction. So a diode works as a one-way valve.

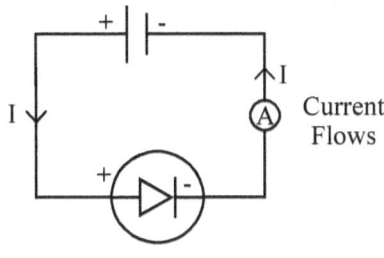

Forward Bias

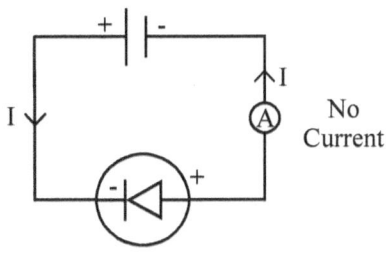

Reverse Bias

Positive of the diode is connected to the positive terminal of the battery, and the negative of the diode is connected to the negative terminal.

Positive of the diode is connected to the negative terminal of the battery, and the negative of the diode is connected to the positive terminal.

Light emitting diodes give out light when a current passes through them. They also allow current to pass in one direction only.

A diode is a rectifier. When it is connected in a circuit with alternating current, then it produces direct current.

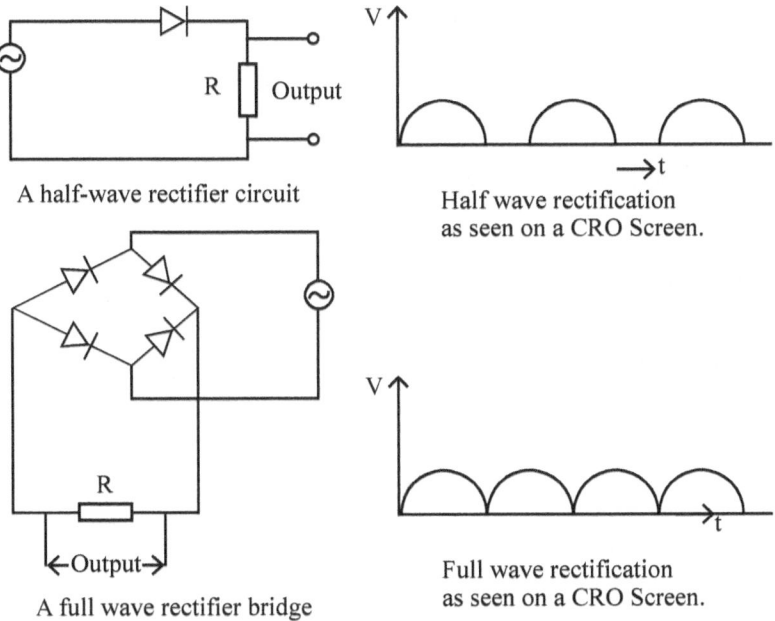

A half-wave rectifier circuit

Half wave rectification as seen on a CRO Screen.

A full wave rectifier bridge

Full wave rectification as seen on a CRO Screen.

CRO $\Rightarrow$ Cathode Ray Oscilloscope

Y axis $\equiv$ volts/cm, X-axis $\equiv$ time/cm.

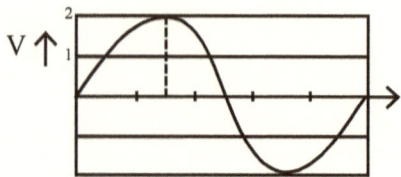

Amplitude = 2 cm. If Y-sensitivity = 2 volts/cm, then the voltage = 4V. Suppose 1 complete cycle = 5 cm horizontally and the X-sensitivity or time base, as it is normally called, is 2 ms/cm, the time period of the wave = 5 × 2 ms = 10 ms.

$T \Rightarrow 10 \times 10^{-3}$ s = 10^{-2} s. frequency f = 1/T = $1/10^{-2}$ = 100 H$_z$.

Paying for Electricity

The cost of electricity consumption is calculated from the amount of 'units' of electricity being used.

1 unit $\Rightarrow$ 1 kWh = 1kW × 1h.
Cost $\Rightarrow$ total units used × price of one unit

The power must be in kW, and the time must be in hours to work out the units.

There is a fixed quarterly charge and VAT has to be added to the total cost at the appropriate rate.

Learning Zone

Electrons are the charge carriers. Charges flow to set up a current. Current is the number of coulombs flowing per second.

Voltage is the amount of energy transferred by each coulomb of charge.

Resistance = voltage ÷ current.

$E = VQ$, $I = Q/t$, $V = E/Q$, $R \Rightarrow V/I = (E/Q)/(Q/t) = Et/Q^2$

In a direct current, the voltage is fixed, and the current passes in one direction only.

In an alternating current, the voltage and the current fluctuate between positive and negative values at a certain frequency. In the UK, the working voltage for our households is 240V and the frequency of the AC voltage is 50 H_z.

Power stations produce electricity, and the power is distributed throughout the country by using transformers, pylons, and cables. This is called the national grid.

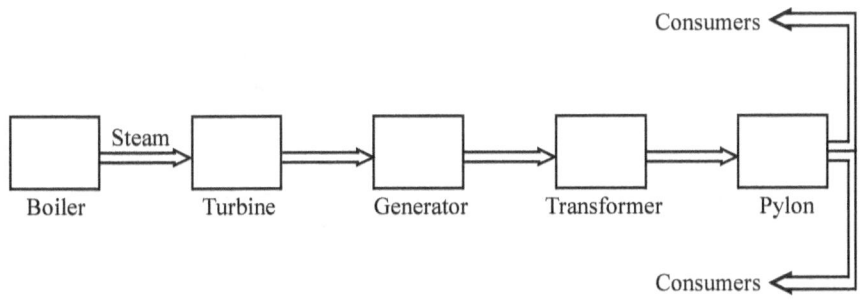

Boiler   ⇒ uses fuel to produce steam ⇒ steam turns the turbine.
Turbine   ⇒ turbine operates the generator.
Generator   ⇒ produces electricity.
Transformer   ⇒ transformers change voltage up or down.
Pylon   ⇒ pylon and the cables distribute power to consumers.

In distributing power, it is desirable that minimum power is wasted during transmission. Power loss is given by the term I^2/R. Since resistance of the cable is constant, the power loss is minimised by increasing the voltage by using step-up transformers, thereby decreasing the current in the grid.

Section 4

Magnetic Fields (1)

Concept Mapping

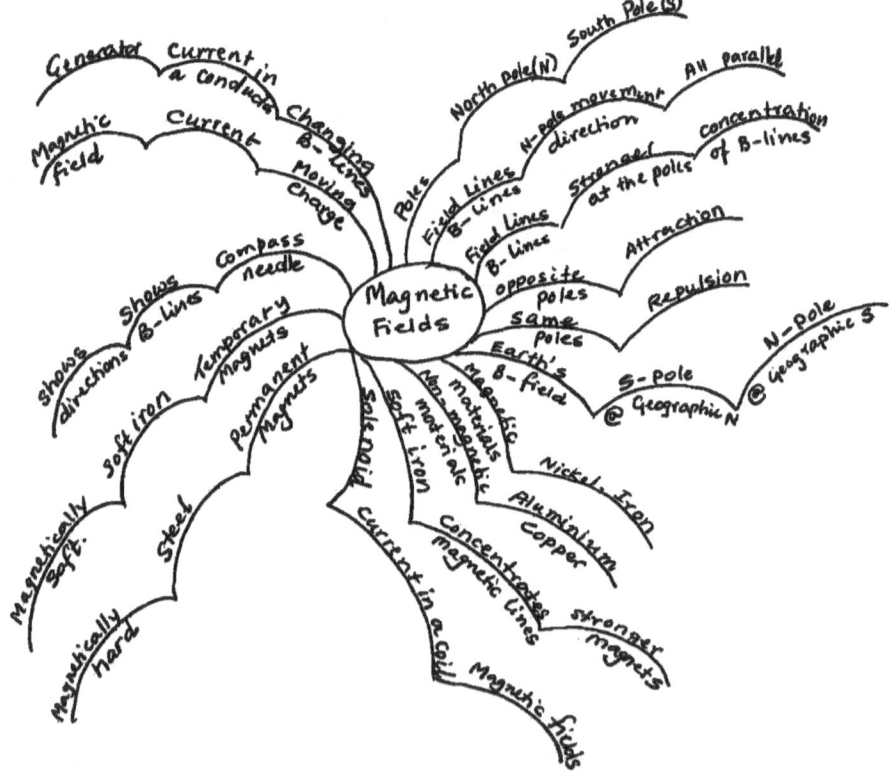

Magnetism

Magnets have poles north (N) and south (S)
Same poles repel, different poles attract. Magnetic field lines (B-lines)
start at the N-pole and end at the S-pole.

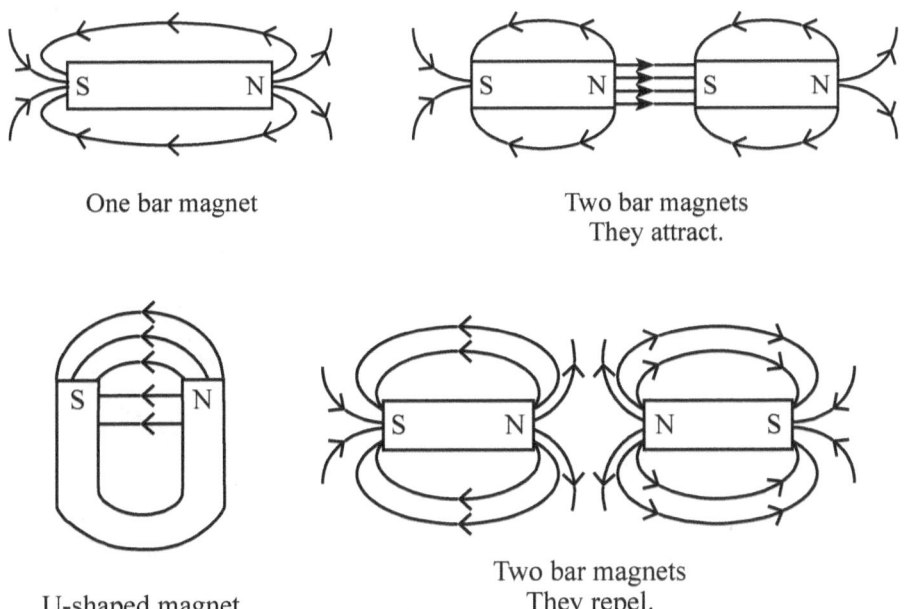

One bar magnet

Two bar magnets
They attract.

U-shaped magnet

Two bar magnets
They repel.

When a bar magnet is suspended freely with a string, it will settle in the geographic north-south direction.

When a magnet is brought near another magnet, they will either attract or repel according to the pole positions.

If a magnetic material is brought near a magnet, they will attract.
If a non-magnetic material is brought near a magnet, there will be no reaction.

Magnetic field lines (B-lines) are called magnetic flux.
When a magnet is moved, the flux lines move too. This is change of flux with time.

Earth is a huge magnet. Its North Pole is at the geographic south and the South Pole is at the geographic north.

The B-lines are the loci of a free N-pole, which moves from the N-pole of the magnet to the S-pole.

Electromagnetism

Concept Mapping

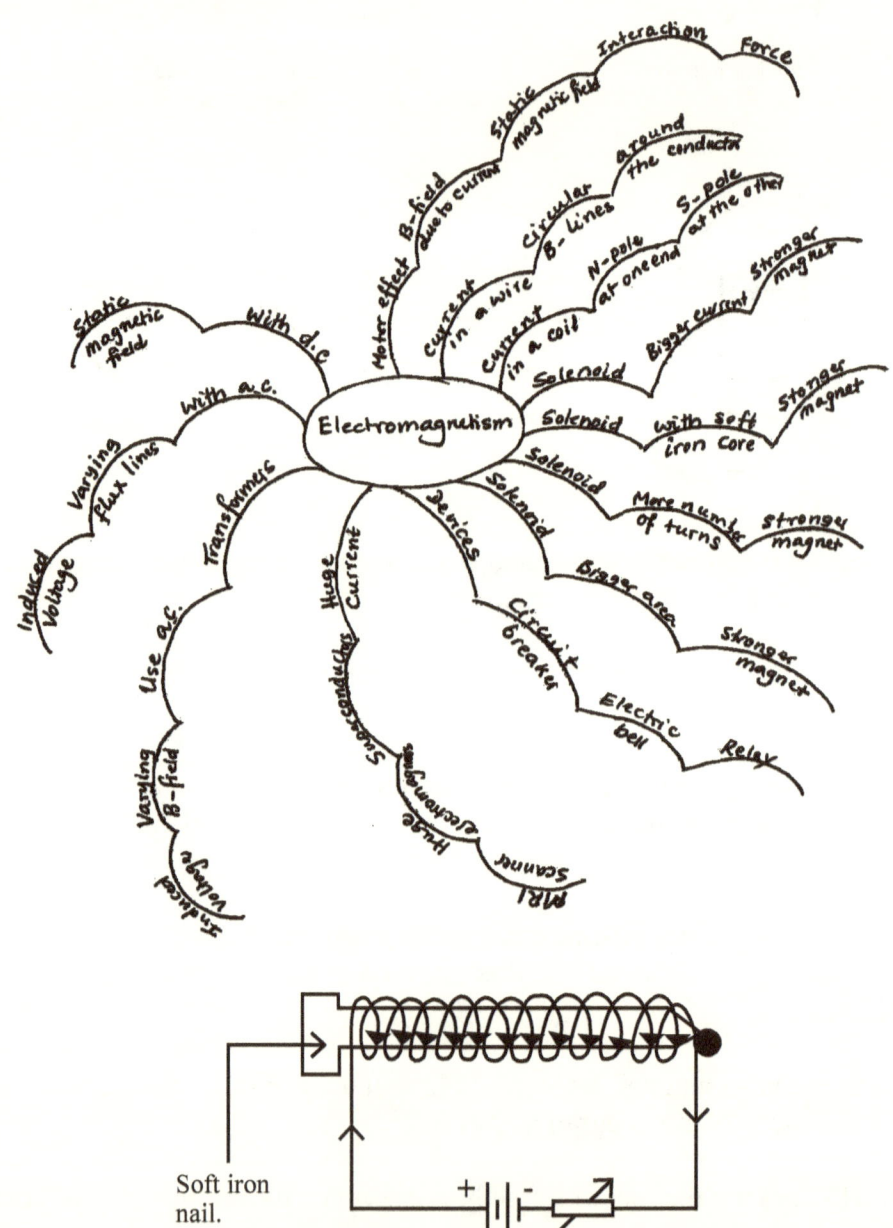

Soft iron
nail.

When a current passes through the coil, the nail becomes a magnet. Therefore, it must have poles. By holding the positively connected end of the coil and looking end on, we can determine the poles.

Clockwise ⇒ south pole   Anticlockwise ⇒ north pole

When the current is switched off, magnetism disappears. Soft iron is easy to magnetise and demagnetise.

Steel is difficult to magnetise and demagnetise. However, once it is magnetised, it becomes a permanent magnet.

Solenoid Properties

A solenoid is a long coil of wire of several turns carrying a current. The coil may or may not have a soft-iron core.

Bigger current	⇒ stronger magnet
More number of turns	⇒ stronger magnet
Bigger area of the coil	⇒ stronger magnet
With soft-iron core	⇒ stronger magnet
Without the soft-iron core	⇒ weaker magnet

B-Lines in a Current-carrying Conductor

Around a current-carrying conductor, the B-lines are circular in a plane perpendicular to the conductor. The direction of these B-lines is given by the 'right-hand grip rule'. Hold your fist with thumb pointing in the direction of the current. The fingers give the direction of the B-lines.

Conventional symbols ⇒ current into the page (×).
⇒ current out of the page (·).

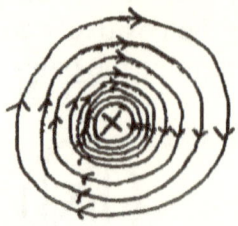
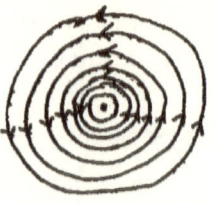

Diagram A shows the circular B-lines due to a conductor in which current is flowing into the page, and diagram B represents B-lines due to a conductor in which current is flowing out of the page.

Whenever there is a current, there is a magnetic field associated with it. Remember, a current is produced by the movement of electric charges. So moving charges cause magnetic fields.

If a conductor carrying a current is placed in a steady magnetic field created between a N-pole and an S-pole, the two magnetic fields interact and produce a force. This force then moves the current-carrying conductor. This is called the motor effect. The direction of motion of the conductor is given by Fleming's left hand rule.

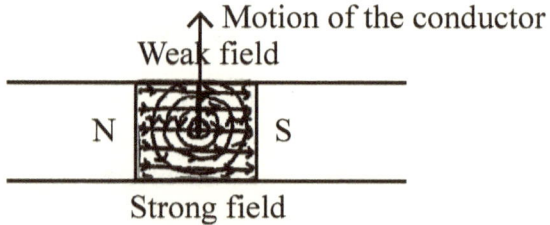

In this figure, the current is flowing out of the page. At the bottom, the B-lines are in the same direction, causing a strong magnetic field. At the top, the B-lines are in the opposite direction, causing weak a magnetic field. The conductor moves up from the strong to the weak magnetic field.

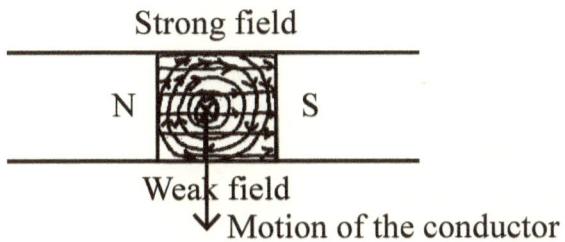

Strong field

N   S

Weak field

Motion of the conductor

In this case, the current is flowing into the page. At the top, the B-lines are in the same direction, causing a strong magnetic field. At the bottom, the B-lines are in the opposite direction, causing a weak magnetic field. As a result, the conductor moves down from strong to weak magnetic field.

Fleming's Left Hand Rule

When the forefinger, middle finger, and the thumb of your left hand are held at right angles to each other,

Fore finger gives magnetic field direction N→S
Middle finger gives current direction, positive → negative

Thumb gives the direction of the moving conductor, i.e. the direction of the force.

The Electric Motor

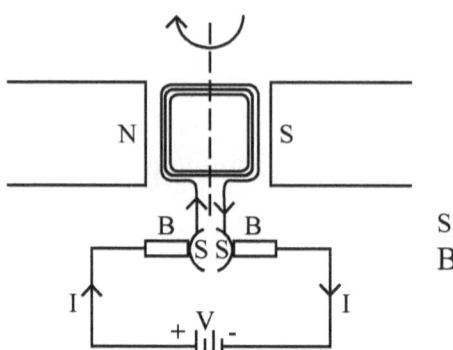

S, S ⇒ split ring commutators
B, B ⇒ Carbon brushes

A coil of several turns is placed in a magnetic field. Each end of the coil is connected to a split ring. Each split ring is touched but not connected by carbon brushes. They are then connected to a battery. The current is moving up the left-hand side and down the right-hand side of the rectangular coil. Apply Fleming's left hand rule to the right-hand side of coil. The right-hand side will come up and the left-hand side will go down, causing a rotation of the coil.

Stronger force will be created, and hence, faster speed of the coil will be achieved by using:

- A bigger current
- More turns in the coil
- A stronger magnetic field

Because of the DC supply, the motor will run in one direction only. By swapping the poles of the magnet or by reversing the polarity of the supply voltage, the movement of the motor can be reversed.

Electromagnetic Induction

Concept Mapping

Changing Magnetic Flux

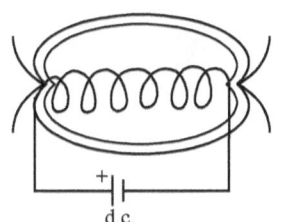

In the coil, there is no change of magnetic flux because of the DC supply. There is a steady B-field.

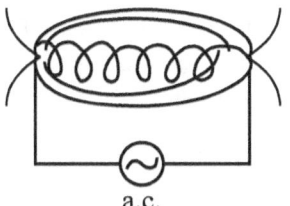

As the supply voltage is alternating, the magnetic flux pattern will alternate at the same frequency. In this case, there is a changing magnetic field.

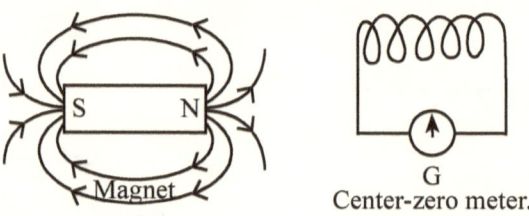

G
Center-zero meter.

The coil or the magnet can be moved, and the flux linkage with the coil can be changed with time. This rate of change of flux linking the coil induces voltage in the coil and the meter will show the induced voltage and its direction.

No movement of the coil or the magnet ⇒ no induced voltage.
The induced voltage (induced emf as it is often called) depends on the rate of change of magnetic flux.

Stronger magnet   ⇒ bigger induced voltage
Bigger coil area   ⇒ bigger induced voltage
More number of turns in the coil   ⇒ bigger induced voltage
Faster movement   ⇒ bigger induced voltage

This is the principle of generation of electricity. Rotation of a magnet inside a coil (as in a bicycle dynamo) or of a coil in a magnetic field (as in a dynamo or an alternator) cause induced voltage in the coil. Remember, DC cannot produce induced voltage.

DC Generator

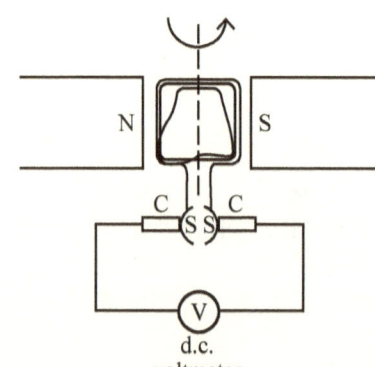

d.c.
voltmeter

S-S ⇒ Split ring communicators
C-C⇒ Carbon brushes

DC ⇒ DC motor in reverse is the DC generator. Rotate the coil by some means, and out comes electricity through the commutator and brush system.

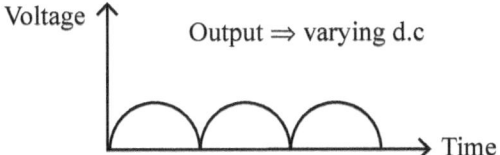

Voltage

Output ⇒ varying d.c

Time

AC Generator

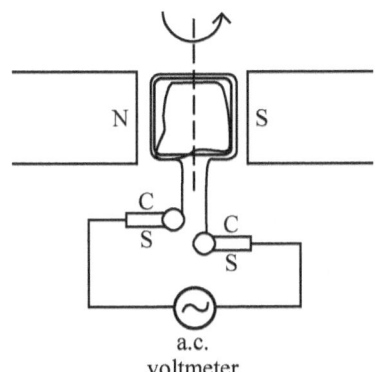

a.c.
voltmeter

S-S ⇒ Slip ring commutators
C-C⇒ Carbon brushes

The only difference in this case is that each end of the coil is connected to a slip ring. The output voltage is an alternating one.

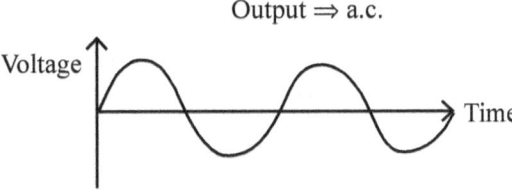

Output ⇒ a.c.

Voltage

Time

Transformers

Concept Mapping

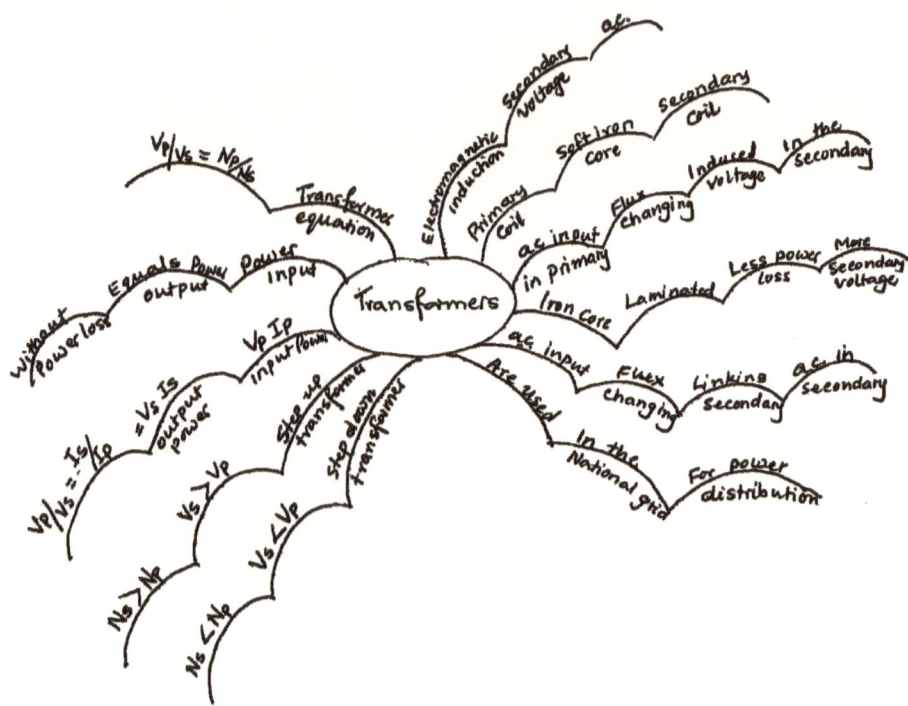

Transformers work on the principle of electromagnetic induction. As the name implies, they transform AC voltages up or down.

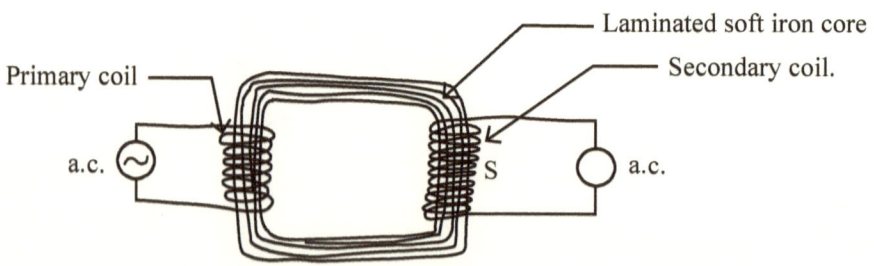

Primary coil

Laminated soft iron core

Secondary coil.

a.c.

S

a.c.

Vp ⇒ Primary voltage
Np ⇒ Primary number of turns
Ip ⇒ Primary current

Vs ⇒ Secondary voltage
Ns ⇒ Secondary number of turns
Is ⇒ Secondary current

Transformer Equations

Vp/Vs = Np/Ns (voltage equation)
Vp Ip = VsIs (power equation)
Percentage efficiency = (VsIs/VpIp) × 100
Step-up transformer ⇒ Vs⟩Vp, Ns⟩Np, Is⟨Ip
Step-down transformer ⇒ Vs⟨Vp, Ns⟨Np, Is⟩Ip

As induced voltage can only be produced when there is a change in magnetic flux, so AC is supplied to the primary coil. Changing magnetic flux is caused by the alternating voltage and they are concentrated in the laminated soft-iron core, so they all pass through the secondary coil. An alternating induced voltage develops in the secondary. So transformers can work with AC only. A DC voltage cannot produce changing magnetic flux in the primary if it is used. Hence it cannot produce induced voltage in the secondary. Therefore, transformers cannot work with DC.

Learning Zone

Alternating current in the primary creates a changing magnetic flux with a certain frequency.

Alternating magnetic flux is inside the soft-iron core.

So alternating magnetic flux is inside the secondary coil.

Changing magnetic flux linking the secondary coil creates the induced voltage in the secondary.

The output voltage is therefore an alternating voltage.

In the electricity transmission process, transformers transfer power from the primary to the secondary coil.

Heat generated in the coils and forming of eddy current loops in the soft-iron core are the main sources of power loss.

Transformers are therefore not 100 per cent efficient.

Power loss (I^2R) in the coils is minimised by using low-resistance copper wires.

Eddy current power loss is minimised by making the core with laminated sheets with insulating material between them.

Section 5

Radioactivity

Concept Mapping

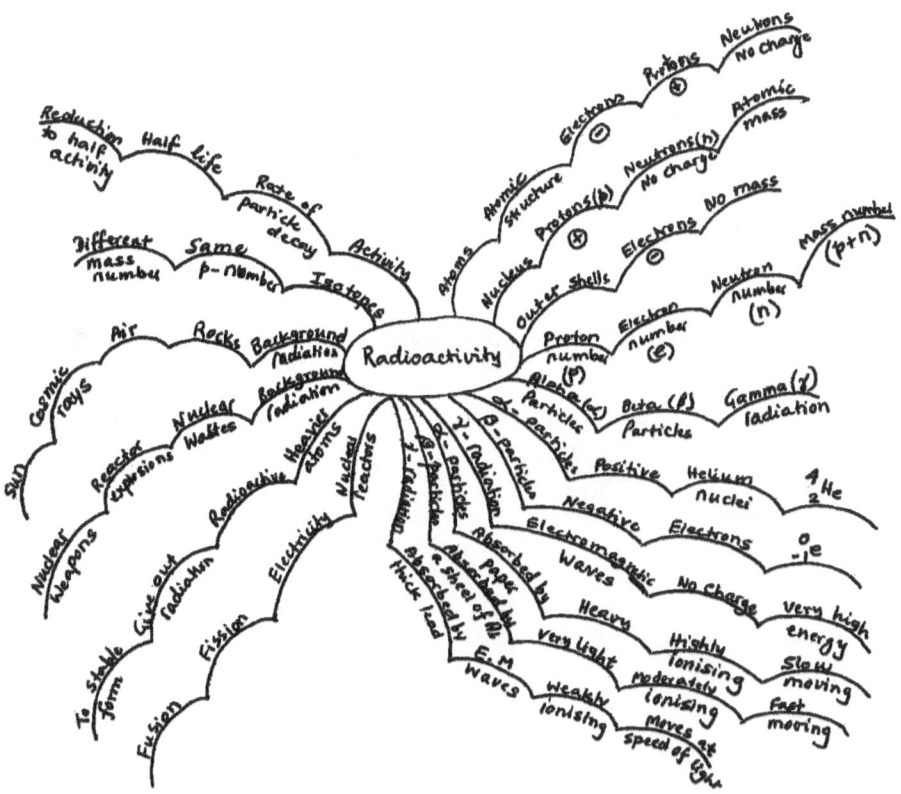

Structure of an Atoms

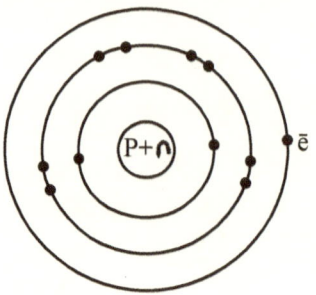

An atom is made up of a central nucleus which contains protons and neutrons. The nucleus contributes to the whole mass of the atom as the electrons are assumed to be of negligible mass. The total number of protons and neutrons gives the mass number (A) of the atom.

$$A = p + n.$$

The electrons move around in different orbits (shells) about the nucleus. Only a specific number of electrons can stay on a specific shell. The first shell can accommodate two electrons only. A filled shell contains maximum of eight electrons. The number of electrons on the outer shell determines the charge of an ion.

In a neutral atom (uncharged atom) electron number equals the proton number because electrons are negative and protons are positive. A neutron has no charge. It only contributes to the mass of the atom.

Particle	Relative mass	Charge	Symbol
Proton	1	+	1p1
Electron	$1/2000 \approx 0$	-1	0e-1
Neutron	1	0	1n0

The structure of all the elements can be found in the periodic table in a chemistry book.

Any atom can be designated as AXZ,

where X   ⇒ an atom (chemical symbol from the periodic table)
A   ⇒ mass number
Z   ⇒ atomic or proton number

So, $Z = p$ and $A = p + n$.
A sodium atom (Na) is designated here as 23Na11

Mass number → 23   p = 11

                   Na   e = 11

Proton number→11   n = 12

Isotopes

Isotopes are elements which have the same proton number but a different
neutron number and hence different mass number, e.g.

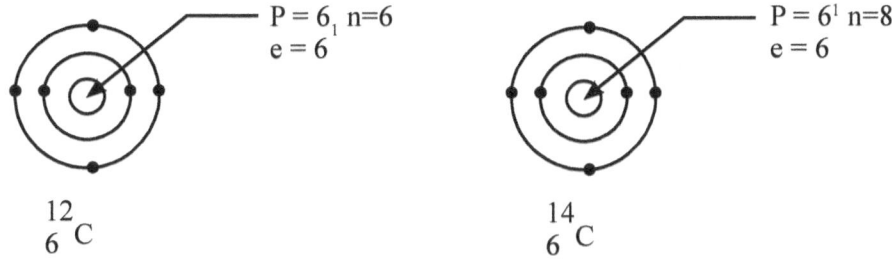

An atom can lose or gain electron to produce an ion. Therefore an ion
is a charged atom. Whenever an atom loses one electron, it becomes a
singly charged positive ion. When it loses two electrons, it becomes a
doubly charged positive ion. Similarly, when an atom gains one electron
it becomes a singly charged negative ion. When it gains two electrons, it
becomes a doubly charged negative ion.

Heavier elements are generally unstable, e.g. 235U92, 226Ra88

They decay by giving off α-particles, β-particles, and γ-radiation to
come to a stable element.

Heavy elements have more neutrons than protons in their nuclei.

    α-particles are positively charged, and they are helium nuclei.
        α ⇒ 4He2 (2p, 2e and 2n).
    β-particles are negatively charged and they are electrons
        β ⇒ 0e-1.

γ-radiation is part of the electromagnetic spectrum. γ-radiation is a wave
and has no charge. It travels at the speed of light ($c = 3 \times 10^8$ m/s).

α-particles are heavier, easily absorbed by a sheet of paper, and highly ionising.

Heavy mass $\Rightarrow$ low speed.

β-particles are very light, almost mass-less, absorbed by a thin sheet of aluminium, and moderately ionising.

Low mass $\Rightarrow$ high speed.

γ-radiation is just a wave carrying very high energy, absorbed by a very thick sheet of lead, and least ionising.

Electromagnetic radiation $\Rightarrow$ speed = speed of light.

Radioactivity in Use

Concept Mapping

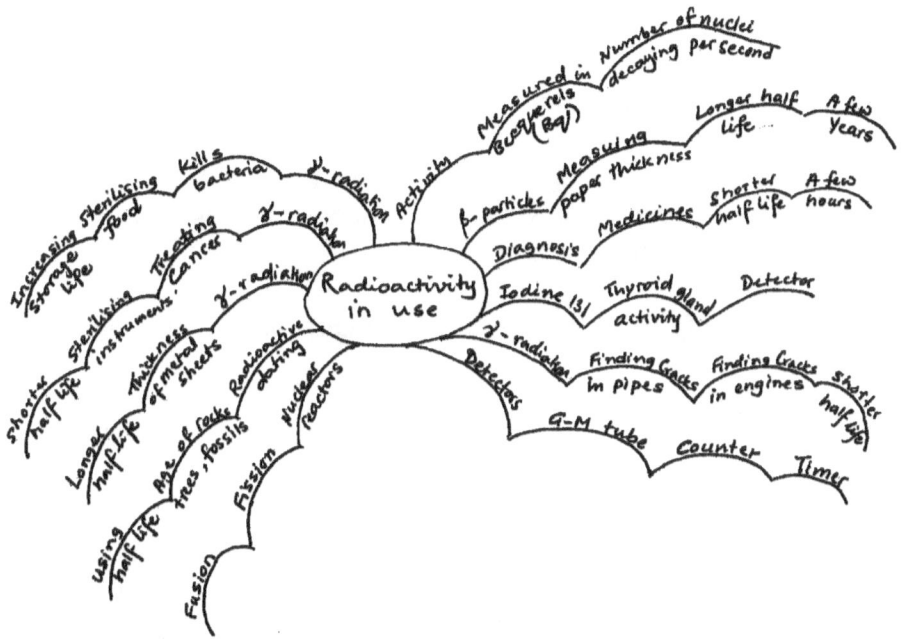

Half-life

Half-life is the time taken by a radioactive element to produce half its original activity measured by the number of radionuclides decaying per second using a G-M tube connected to a counter-timer. The number of radionuclides decaying per second is called a Becquerel (Bq). This is the unit of radioactivity.

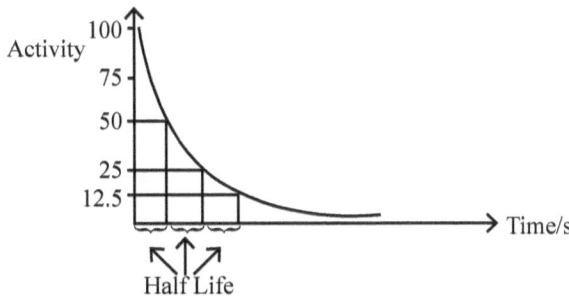

From the graph, it can be seen that
  after 1 half-life, the fraction remaining = ½
  after 2 half-lives, the fraction remaining = ¼

Therefore, after n half-lives, the fraction remaining = $1/n^2$.
If the half-life is fifteen minutes, there will be four half-lives in one
  hour.
If the half-life is four days, there will be seven half-lives in twenty-eight
  days.

For long-time observations, isotopes of longer half-lives are used so that
activity does not reduce considerably.

For manufacturing paper of a certain thickness, half-life of some years
is necessary.

For hospital diagnostic purposes, half-life of some hours should be
considered.

Gamma radiations are used for radiation treatment of cancerous tissues.
It is also used in finding cracks in oil pipes and engines. They kill bacteria
in food and are used in treating food for long-time storage.

α—and β—particles are deflected by electric and magnetic fields
because both of them are charged particles.

γ-radiation is not affected by electric and magnetic fields because the
  radiation is an electromagnetic wave with no charge.

α-decay

  222Ra88 →222Rn86 + 4 He 2 (α)

Because of the removal of an α-particle, the mass number of the parent
nuclide is reduced by 4 and the proton number is reduced by 2 to produce
the daughter nuclide.

β-decay

$218Po84 \rightarrow 218At85 + 0e\text{-}1 \ (\beta)$

Because of the removal of a β-particle, the mass number of parent nuclide remains unchanged and the proton number is increased by 1 to produce the daughter nuclide.

When α and β particles are emitted from a radioisotope by decay processes, then the nucleus is left in an unstable or excited state with excess energy. The nucleus then releases γ-radiation of very short wavelength to come to a stable state.
Therefore γ-radiation originates in the nucleus.

γ-decay

It is a two-stage decay process. Emission of γ-radiation is preceded by α—or β-decay, e.g. in γ-decay of 60Co27,

Stage 1 $60Co27 \rightarrow 60Ni28 + 0e\text{-}1(\beta)$.
β-decay

Stage 2 $60Ni28 \rightarrow 60Ni28 + 0\gamma0$.
γ-decay

In the γ-decay process, no new element is formed.

Learning Zone

Unstable elements are radioactive.
Radioactive elements decay, giving off particles or waves.
During decay process, different elements are produced.
Radioactivity ceases when a stable element is formed.

   α –particles are 4He2, have positive charge, highly ionising.
   α –particles get easily absorbed by a piece of paper.
   β – particles are 0e-1, have negative charge, moderately ionising.
   β – particles get absorbed by a sheet of aluminium.
   γ – radiation is produced after α—or β-decay has taken place.

γ – radiation has no charge or mass. They are electromagnetic waves.

γ – radiation travels at the speed of light (3×10^8 m/s).

γ – radiation gets absorbed by a thick block of lead.

Number of radionuclides decaying per second is called activity.

From any given time, the time taken to produce half the original activity is called a half-life.

β-particles and γ-radiation are mostly used in industries, hospitals, and food processing.

Shorter half-lives are preferred for diagnostic purposes.

Longer half-lives are essential in manufacturing industries.

All radioactive elements and their decay products are harmful to animals and plants.

Section 6

Energy and Energy Resources

Concept Mapping

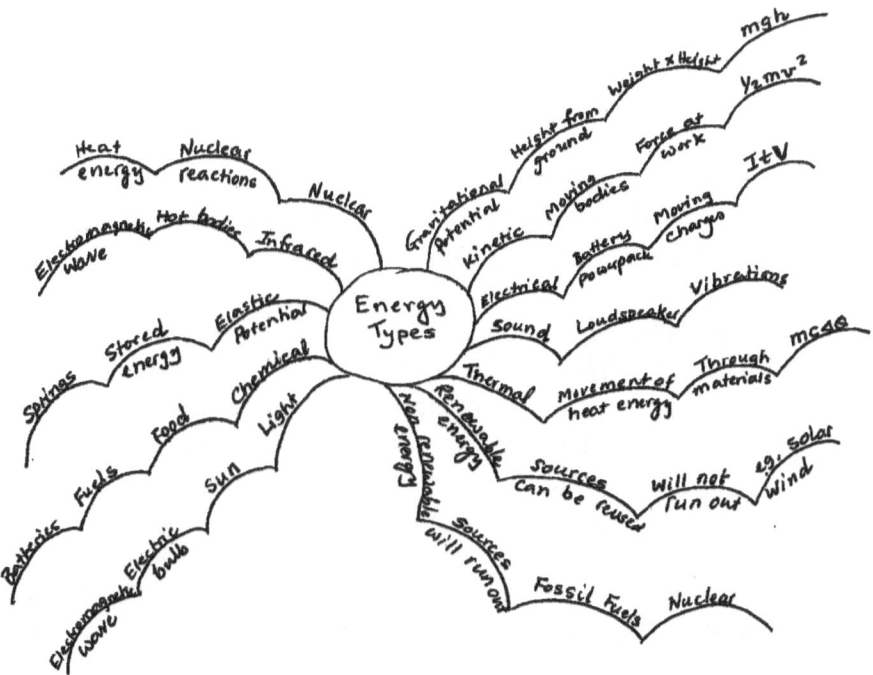

Energy

Where there is energy, there will be work done in certain form. Energy can change forms but can never be destroyed.

Energy ≡ work

Gravitational potential energy is the work done when an object is moved up the ground to a certain height. The work is done against the gravitational pull of the earth.

Kinetic energy is associated with moving bodies. The movement is caused by a force. For a falling object, its gpe decreases and KE increases.

Electrical energy is the result of the movement of charge in a conductor. The movement is caused by the electrical push on the free electrons from the chemical energy in the battery.

Sound energy is produced by the vibration of a medium. The vibration is caused by an external force. The compressions and rarefactions of the particles in the medium produce the wave with a certain amplitude and frequency.

Light energy is an electromagnetic wave. Light comes from the sun or is produced in an electric bulb. Light is produced by the combined vibrations of electric and magnetic fields associated with it.

Nuclear energy is the result of nuclear reactions of radioactive elements, e.g. uranium and plutonium. In a fission reaction, a heavy element breaks down into smaller elements, producing a large amount of heat energy. In a fusion process, smaller elements combine to produce a heavier element under very high temperature and pressure. Fusion reaction produces much more heat energy than fission.

Infrared heat radiation is an electromagnetic wave which gives the sensation of heat. It can travel through vacuum without any medium.

Elastic potential energy is a type of stored energy due to a force. Springs, elastic bands, and bow and arrow possess elastic potential energy.

Chemical energy is also a type of stored energy possessed by fuels, batteries, etc. Chemical energy from food provides us with energy to move, work, and play.

Thermal energy is a name given to heat energy that moves through materials by conduction and convection. Again, a force is at work for electron flow and vibrations of atoms in conduction and movement of particles of the medium in the convection process. Conduction takes place in solid conductors, and convection takes place in liquids and gases.

Energy Resources

Concept Mapping

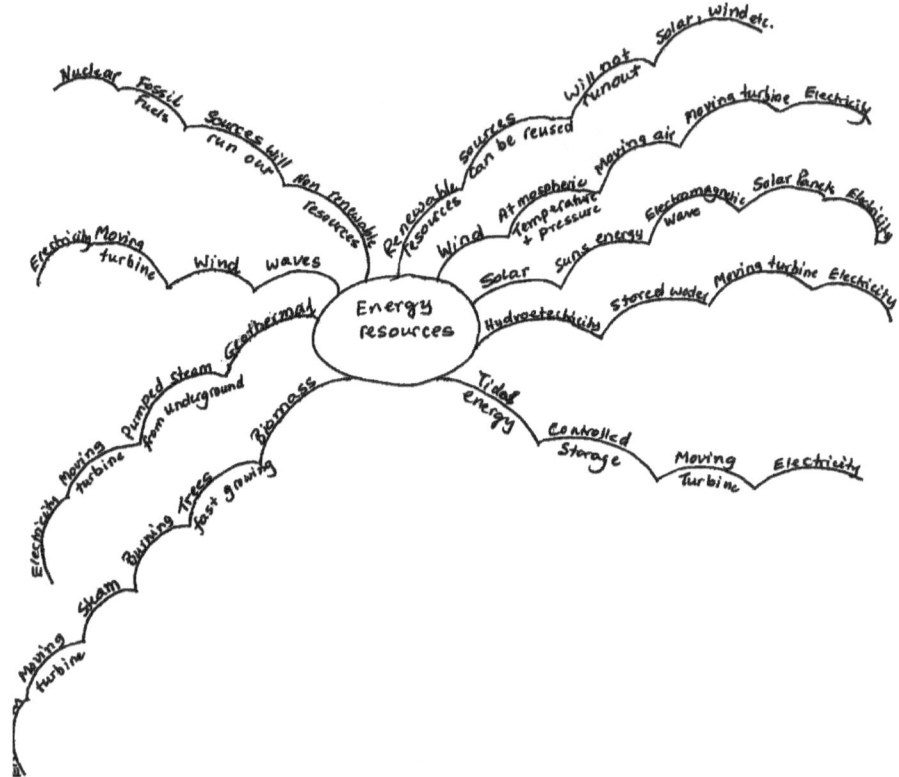

The interior of the sun is extremely hot. The temperature there is about 7000°C. The sun produces its energy by fusion of hydrogen nuclei to form helium nuclei. This electromagnetic wave forms our visible light and infrared radiation.

Non-renewable energy resources like coal, oil, gas, and nuclear fuels will run out one day.

The world economy is highly dependent on energy consumption. Renewable energy resources that can be used over and over again are vitally important and necessary.

Wind energy $\Rightarrow$ movement of air $\Rightarrow$ high pressure to low pressure

Solar energy $\Rightarrow$ Sun's energy $\Rightarrow$ converted by solar panels $\Rightarrow$ electrical

Tidal energy $\Rightarrow$ stopped tides $\Rightarrow$ moving turbine $\Rightarrow$ electricity

Wave energy $\Rightarrow$ winds produce waves $\Rightarrow$ moving turbine $\Rightarrow$ electricity

Geothermal energy $\Rightarrow$ pumped out steam from underground hot rocks $\Rightarrow$ moving turbine $\Rightarrow$ electricity

Biomass energy $\Rightarrow$ burning fast growing trees to produce steam $\Rightarrow$ moving turbine $\Rightarrow$ electricity

Hydroelectric energy $\Rightarrow$ stored water in higher grounds $\Rightarrow$ moving turbine $\Rightarrow$ electricity.

It is to be remembered that it is always a chain process of energy transfer from one form to another which produces electrical energy in the end.

Undesirable effects of energy production

Fossil fuels $\Rightarrow$ coal, oil, and gas $\Rightarrow$ CO_2, SO_2 production $\Rightarrow$ global warming

Nuclear fuels $\Rightarrow$ harmful wastes $\Rightarrow$ catastrophe $\Rightarrow$ Chernobyl, Fukushima

Wind power $\Rightarrow$ many wind turbines $\Rightarrow$ ugly landscape

Solar power $\Rightarrow$ clear sky $\Rightarrow$ many solar panels

Tidal power $\Rightarrow$ high initial cost

Wave power $\Rightarrow$ wind-controlled $\Rightarrow$ high initial cost

Geothermal $\Rightarrow$ high initial cost

Biomass $\Rightarrow$ burning wood $\Rightarrow$ CO_2 $\Rightarrow$ reduction by replanting

Hydroelectric $\Rightarrow$ higher initial cost $\Rightarrow$ controlled energy distribution.

The Sun and the Energy Resources

Concept Mapping

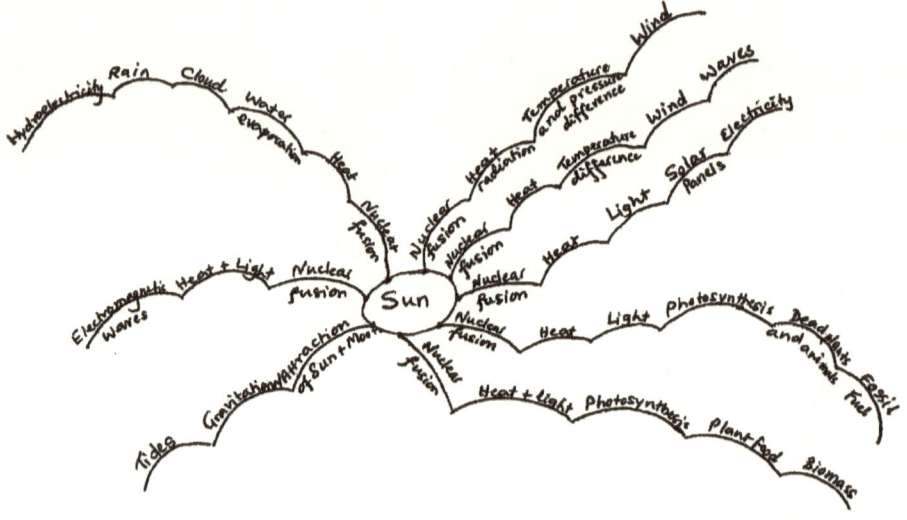

The sun is one of the millions of stars in our Milky Way galaxy. It is conceivable that the sun is the source of all different energy resources. Nuclear reactions in the sun produce heat and light.

The relationship between the sun and the different energy resources can be summarised as follows:

Heat ⇒ temperature difference ⇒ moving air ⇒ wind
Heat ⇒ temperature difference ⇒ wind ⇒ waves
Heat ⇒ water evaporation ⇒ cloud ⇒ rain ⇒ hydroelectricity
Light ⇒ photosynthesis ⇒ plants decay ⇒ fossil fuel
Light ⇒ photosynthesis ⇒ plants ⇒ biomass
Light ⇒ photosynthesis ⇒ animal food ⇒ decay ⇒ fossil fuel
Light ⇒ solar panels ⇒ electricity
Gravitational attraction ⇒ sun + moon ⇒ tides.

Section 7

Glimpses into Outer Space

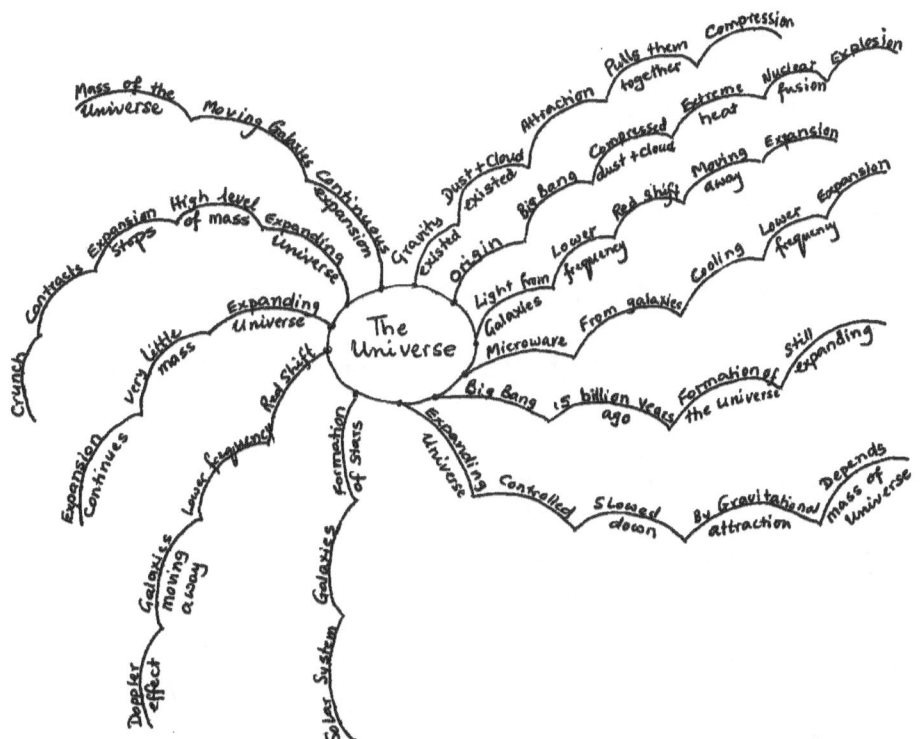

A Few Steps to Big Bang

Existence of gravity, dust and cloud is accepted.
Gravitational attraction pulls dust and clouds.
Dust and cloud are highly compressed.
Compression results in extremely high temperature.
Nuclear fusion takes place resulting in an explosion.
That is the Big Bang and the birth of the universe.

Telltale Sign of the Expanding Universe

Frequencies of light from distant galaxies are measured using radio spectroscopy.

They are found to have lower frequencies than expected.

This proves that the light is shifted toward red of the electromagnetic spectrum.

This is called the red shift or Doppler effect.

This proves that the galaxies are moving away from their original position.

Hence the universe is expanding.

Another argument is based on the heating effect of microwaves.

Microwaves from distant galaxies have been found to be of lower temperature than expected.

This is called microwave cooling.
This also gives lower frequencies of the microwave than expected.
This shows that the galaxies are moving away.
This also proves that the universe is expanding.

To Expand or Not to Expand

This depends on the total mass of the universe.

Gravitational attraction opposes the universe expanding.
If there is a high level of mass, expansion will stop, the universe will contract, and there will be a 'big crunch'.

If there is very little mass, gravitational attraction will be very less and the universe will continue to expand.

The Solar System

Concept Mapping

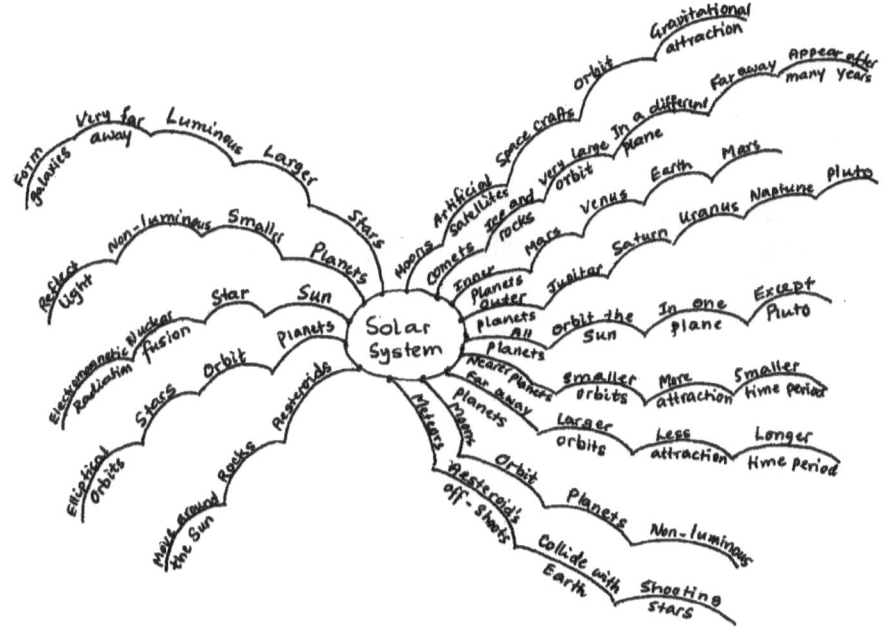

Gravitational attraction controls the solar system.

Planets, moons, asteroids, and comets have orbits.

Orbits of Pluto and the comets are in different planes than the rest.

Nearer planets have smaller orbits, faster speeds, and smaller time periods.

Faraway planets have larger orbits, slower speeds, and longer time periods.

Planets are non-luminous. They reflect the sun's light.

Stars are very large and luminous. The sun is a star.

The Milky Way galaxy is made up of billions of stars.

Our sun is just one of these stars in the Milky Way galaxy.

Billions and billions of such galaxies comprise the universe.

All planets have elliptical orbits around the sun.

Moons move around planets.

Gravitational force controls the radius of orbit, speed, and time period of the orbits of the moving bodies in the solar system.

Revisiting Newton

Concept Mapping

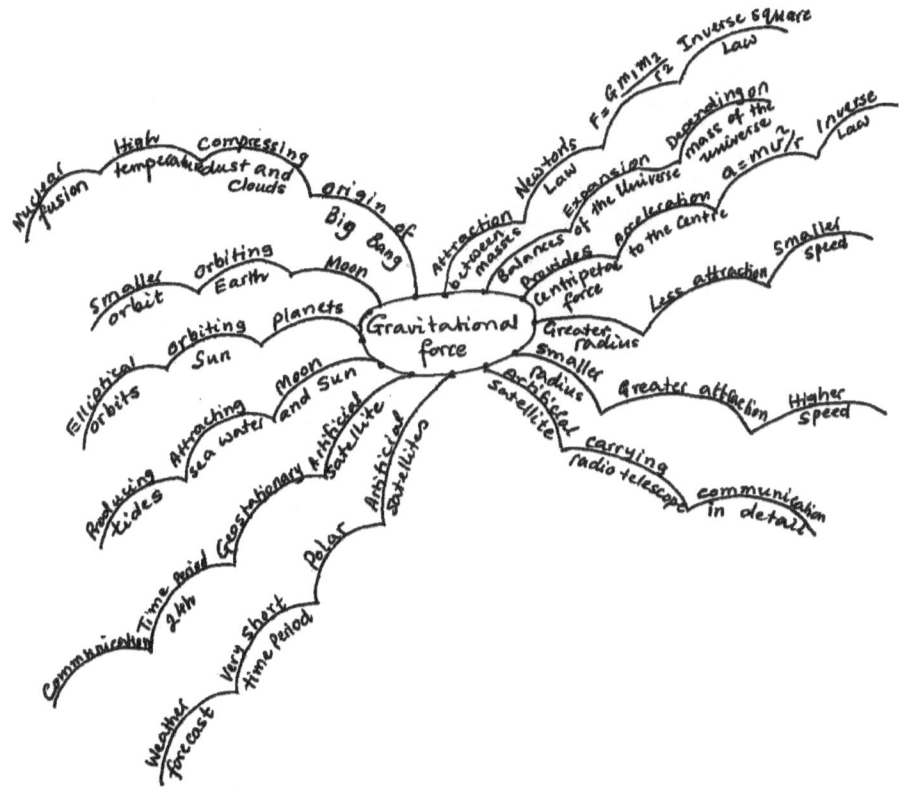

This force is an attraction between masses.

Newton's force between two masses m_1 and m_2 with a separation r between the centres of m_1 and m_2 is given by

$F_N = G\,m_1 m_2\,/\,r^2$ where G = universal gravitational constant

$F_N \propto 1/r^2 \Rightarrow$ inverse-square law

This gravitational force causes the planets, moons, comets, asteroids, and artificial satellites to have orbits of certain radii and moving with a certain speed.

Artificial satellites can carry radio telescopes to provide detailed information of surfaces of planets.

Geostationary satellites have T = 24h and are used for communication.

Polar satellites have a very short time period of some hours and are mainly used for forecasting weather.

Newton's force equals the centripetal force when a body of mass m moves in a circular orbit of radius r with a constant velocity v.

Centripetal force = m v^2/r acting toward the centre.

Compare   F = m a

Centripetal acceleration = v^2/r acting toward the centre.

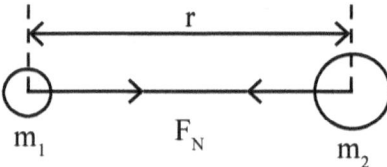

Newton's force F_N = G $m_1 m_2/r^2$, attraction between two masses.

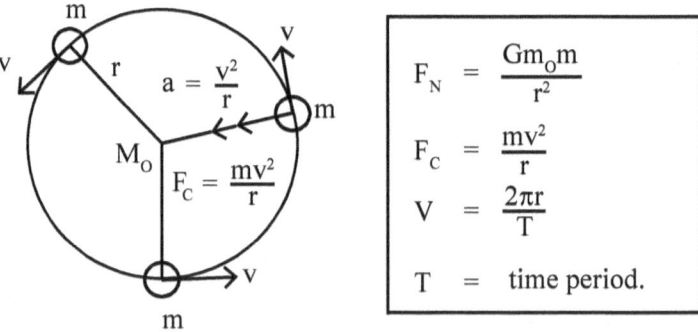

Mass m is moving with a constant velocity v in a circular orbit of radius r with mass m_0 at the centre.

Centripetal force is $F_C = mv^2/r$ and centripetal acceleration is v^2/r.

Considering a circular orbit for a planet, $F_N = F_C$.

Equating $G\, m_o m/r^2 = mv^2/r$

$\Rightarrow v^2 = G\, m_o/r$

$\Rightarrow 4\pi^2 r^2/T^2 = Gm_o/r$ ($v = 2\pi r/T$, T = time period)

$\Rightarrow T^2/r^3 = 4\pi^2/GM$ (constant)

$\Rightarrow T^2 = (4\pi^2/GM)\, r^3$

$\therefore T^2 \propto r^3$.

The square of the time period is directly proportional to the cube of the radius of the orbit of the planet.

T^2/r^3 = constant for a planet

PART II

Physics for Upper Stages

Section 8

Gravitational Field

Concept Mapping

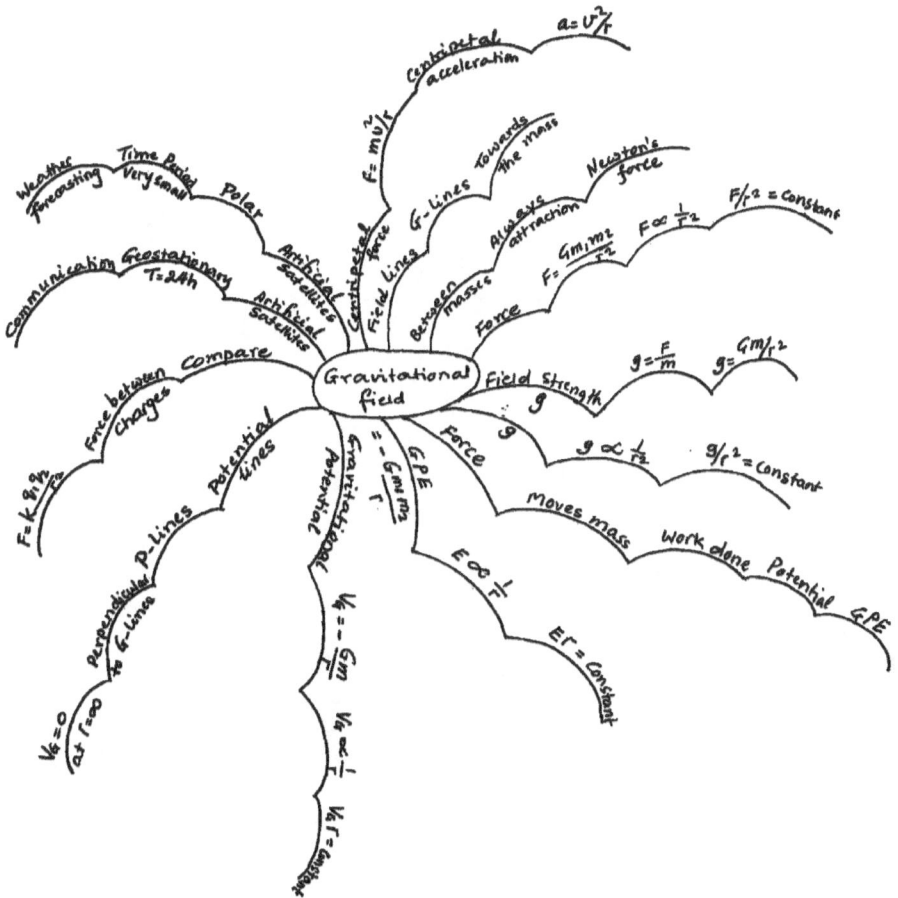

Gravitational field is where the force of attraction is present. Field lines (designated here as G-lines) are toward the mass.

Potential lines (designated here as P-lines), or potential surfaces (3D), are perpendicular to the G-lines.

Gravitational potential is the amount of work done by a mass of 1 kg when it is brought from infinity to a point where the potential in measured. The unit for gravitational potential is Jkg^{-1}.

Gravitational potential energy refers to work done by any mass except 1 kg. The unit is J.

Gravitational field strength is the gravitational pull of the earth on a mass of 1 kg.

Gravitational field lines (G-lines)

One solitary mass
G – lines

- Radial
- Uniform
- In 3D space.

Gravitational potential line (P-lines)

One solitary
mass
P – lines
P – surfaces
(3D)

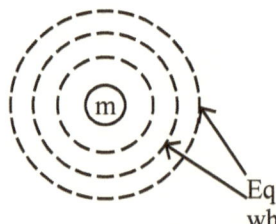

Equipotential Surfaces (3D)
where the potential is same at any point
on respective surfaces.

- P – lines are perpendicular to the G – lines.

Gravitational potential

Gravitational potential
(mass = 1 kg)

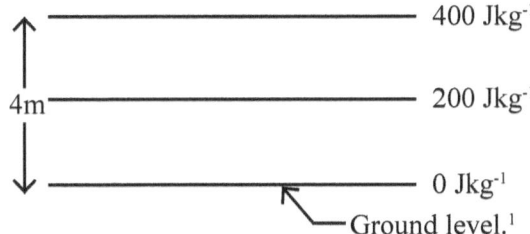

4m

400 Jkg^{-1}

200 Jkg^{-1}

0 Jkg^{-1}
Ground level.[1]

- Potential depends on difference in height only

Gravitational potential energy (gpe)

Gravitational potential
energy (gpe) (mass =
5 kg)

4m

2000J

1000J

0 J
Ground level.[1]

- Potential energy refer to any mass except 1 kg

Gravitational force

Formula $F = G\, m_1 m_2 / r^2 \Rightarrow Fr^2 = $ constant
Proportionality $F \propto 1/r^2 \Rightarrow$ inverse-square law
Graph

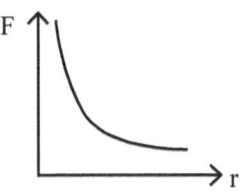

Information
from graph

Area under the graph
= work done

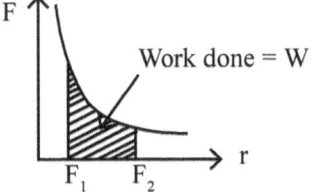

Gravitational field strength	Formula $g = Gm/r^2 \Rightarrow gr^2 = $ constant Proportionality $g \propto \alpha 1/r^2 \Rightarrow$ inverse-square law Graph

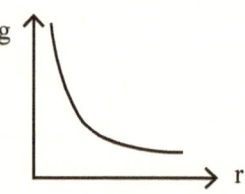

Information from graph   Area under the graph = |potential|

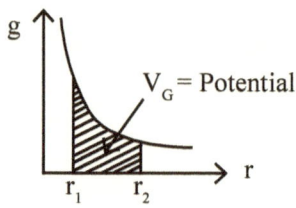

Gravitational potential	Formula $V_G = -Gm/r$. $V_G\, r = $ constant Proportionality $V_G \propto 1/r \Rightarrow$ inverse proportionality Graph

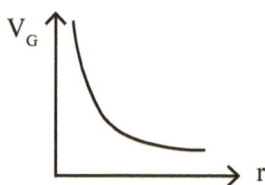

Information from graph   Gradient at a certain point (r_0) = field strength at that point

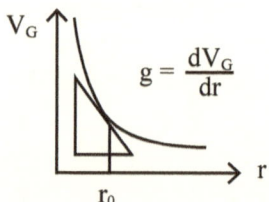

Gravitational
potential

Formula gpe $= - Gm_1m_2/r$, gpe $\times r =$ constant
energy
Proportionality gpe $\propto 1/r \Rightarrow$ inverse
proportionality
Graph

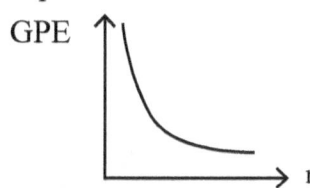

Information   Gradient at a certain point (r_o)
from graph $=|$ gravitational force $|$ at that point.

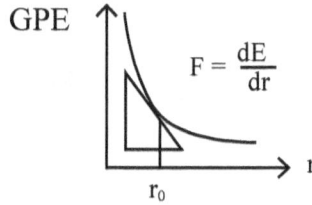

Bodies Moving in a Circle

Concept Mapping

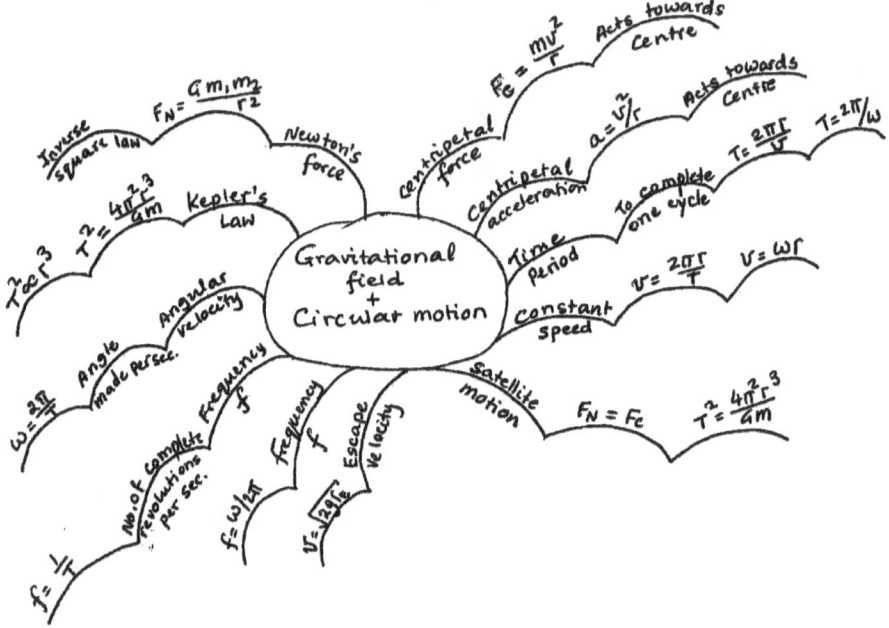

When an object moves in a circle of radius r with a constant velocity v, then the direction of the velocity vector changes on each point of the orbit.

Therefore, this results in the centripetal acceleration acting toward the centre. Hence, a centripetal force develops, acting toward the centre.

Centripetal force $Fc = mv^2/r = mr\omega^2$.
Centripetal acceleration $a = v^2/r = r\omega^2$.
Angular velocity ω = angle swept per sec = $2\pi/T$.
Time period = T = time for one complete revolution.
Linear velocity $v = 2\pi r/T \Rightarrow v = \omega r$.
Frequency = number of complete revolution per second $\Rightarrow f = 1/T$.

Consider an artificial satellite of mass m_s moving with a velocity v in a circular orbit of radius r having a mass m at the centre of the circle.

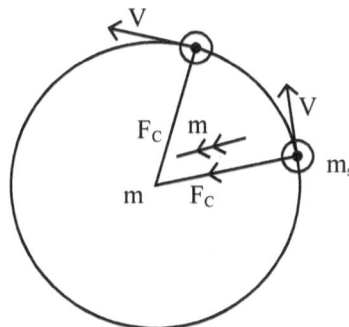

F_N between m and m_s is given by $F_N = Gm\,m_s/r^2$.
F_C for mass m_s is given by $F_C = m_s v^2/r$.
For m_s to move in this orbit with a velocity v and radius r,

$$F_N = F_C.$$
$$\Rightarrow G\,mm_s/r^2 = m_s v^2/r$$
$$\Rightarrow V^2 = Gm/r$$
$$\Rightarrow 4\pi^2 r^2/T^2 = Gm/r$$
$$\Rightarrow T^2 = 4\pi^2 r^3/Gm$$
$$\therefore\ T^2/\,r^3 = (4\pi^2/Gm) = \text{a constant.}$$

This is Kepler's law for the motion of a planet.
Radian measure is useful when dealing with circular motion and waves.

π radian = 180° $\Rightarrow$ ½ revolution
$\pi/2$ radian = 90° $\Rightarrow$ ¼ revolution
$3\pi/2$ radian = 270° $\Rightarrow$ 3/4 revolution
2π radian = 360° $\Rightarrow$ 1 complete revolution
$G = 6.67 \times 10^{-11}$ N m² kg⁻² (universal gravitational constant)
$M_E = 6.0 \times 10^{24}$ kg   (mass of the earth)
$r_E = 6.4 \times 10^6$ m   (radius of the earth).

If a mass m_1 on the surface of the earth can be given a velocity which will move it to infinity, we call the velocity escape velocity.
The potential energy possessed by m_1 at a distance r_E from the earth of mass M_E is given by
$$E = -\,GM_E m_1/r_E.$$

Mass m_1 must have this amount of KE to escape from the gravitational field.

$\therefore$ KE of $m_1 \geq GM_E m_1 / r_E$
$\Rightarrow$ ½ $m_1 v^2 \geq GM_E m_1 / r_E$
$\Rightarrow$ $v^2 \geq 2GM_E / r_E$
$\Rightarrow$ $v^2 \geq 2gr_E$ [$g = GM_E / r_E^2$]
$\therefore$ $v = \sqrt{(2gr_E)}$

Putting

g $= 9.8$ N Kg^{-1}
r_E $= 6.4 \times 10^6$ m
v $= 11.2 \times 10^3$ ms^{-1}

Therefore the minimum velocity required for an object to leave the surface of the earth is about 11 km s^{-1}.

Section 9

Electric Field

Concept Mapping

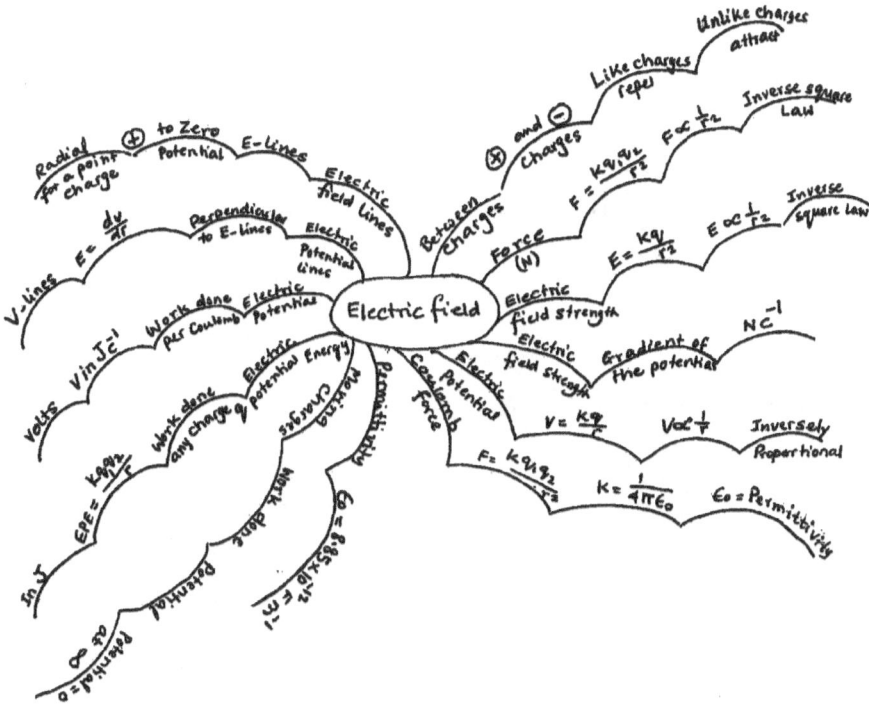

There are two types of electric charge: positive (+) and negative (-). Electric field is where the force can act on a charge.

Coulomb force between two charges q_1 and q_2 with a separation of r between them is given by

$F = k \, q_1 q_2 / r^2$ (N) where $k = 1/4\pi\varepsilon_0$.
ε_0 = Permittivity = 8.85×10^{-12} F m^{-1}.

Electric field strength is the force on 1 coulomb.

$E = kq/r^2$ (NC^{-1})

Electric potential refers to the amount of work done by 1 coulomb in moving from infinity to the point where the potential is measured.

$V = kq/r$ (JC^{-1})

Electric potential energy is the amount of work done by any charge except 1 coulomb.

$epe = kq_1q_2/r$ (J)

Electric Field Lines (E-lines)

Electric field lines due to a positive charge +q.

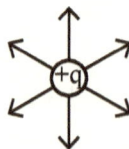

- Radial
- Uniform

Electric field lines due to a negative charge −q

- Radial
- Uniform

Field lines due to one positive and one negative charge.

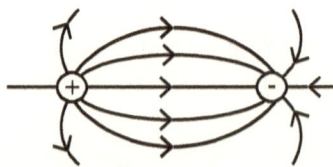

- Non-uniform

Field lines due to two positive charges.

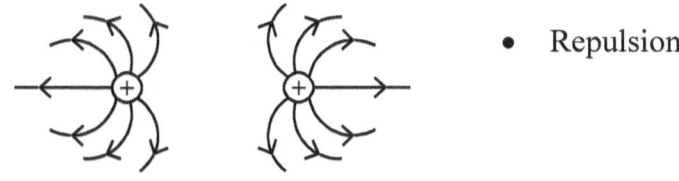

• Repulsion

Field lines due to two negative charges.

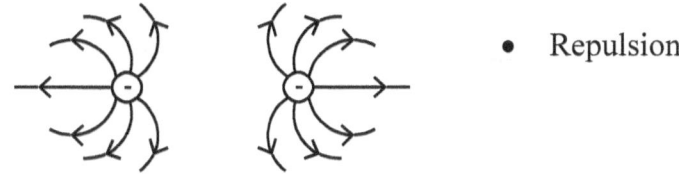

• Repulsion

Electric potential lines due to a positive charge +q.

The potential lines (potential surfaces in 3D) are perpendicular to the E-lines.

Electric potential lines due to a negative charge −q.

The potential lines (potential surfaces in 3D) are perpendicular to the E-lines.

Electric potential between two electrodes.

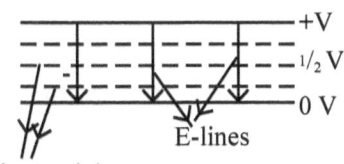

E-lines

Equipotential
Lines

The potential lines (V-lines) are parallel to the electrodes.

The potential lines (V-lines) are parallel to the electrodes.

Electric force   Formula $F = k\, q_1 q_2 / r^2$
                       Proportionality $F \propto 1/r^2 \Rightarrow$ inverse-square law

                     Information from   The area under the graph = work done
                     graph $Fr2$ = constant

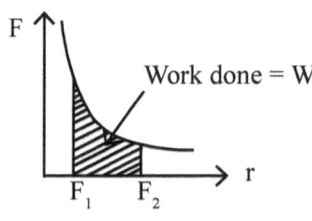

Electric field   Formula $E = kq/r^2$
strength   Proportionality $E \propto 1/r^2 \Rightarrow$ inverse-square law

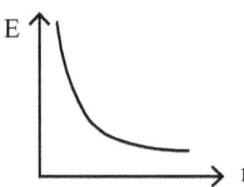

                     Information from   Area under the graph
                     graph   = potential $Er2$ = constant

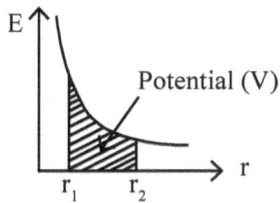

| Electric potential | Formula | $V = kq/r$ |
| | Proportionality | $V \propto 1/r \Rightarrow$ inverse proportionality |

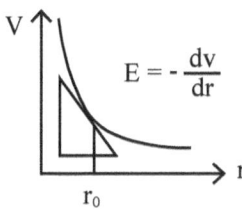

$$E = -\frac{dv}{dr}$$

| | Information from graph | Gradient at a point (r_0) = Field strength |
| | | Vr = constant |

Electric potential energy (EPE)	Formula	$epe = kq_1q_2/r$
	Proportionality	$epe \propto 1/r \Rightarrow$ inverse proportionality
	Graph	

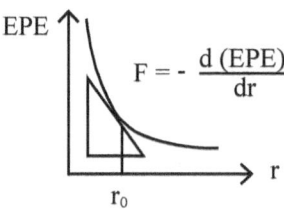

$$F = -\frac{d(EPE)}{dr}$$

| | Information from graph | Gradient at a point (r_0) = Electric force |
| | | $EPEXr$ = constant |

The force between two charges depends on the medium in which the charges are placed. This is where the word permittivity (ε_0) comes in. The medium between two electrode plates is called a dielectric. Air, paper, plastic, and vacuum are all dielectric media. They affect the electric field.

ε_0 = permittivity of the free space.
ε_r = 1 for a vacuum.
ε_r = relative permittivity.

Permittivity of a medium $\varepsilon = \varepsilon_0 \varepsilon_r$.

Learning Zone

Electric force can be attractive and repulsive.
Electric force is between two charges.
A dielectric medium affects the force between charges.
Coulomb's electric force $F = (1/4\pi\varepsilon_0)q_1 q_2/r^2 \Rightarrow$ inverse-square Law.
Electric field strength E = force on 1 coulomb = $F/q = (1/4\pi\varepsilon_0)q/r^2$.
Electric field strength E = $-dV/dr$, where V = electric potential.
Electric potential V = Work done by 1 coulomb = $Er = (1/4\pi\varepsilon_0)q/r$.
Electric potential energy = work done by any charge = $Fr = (1/4\pi\varepsilon_0)$ $q_1 q_2/r$.
Electric field strength = − gradient of the potential at a point.
E = $-dV/dr$ at any point.
F = $-d(epe)/dr$ at any point.
Electron charge = -1.6×10^{-19}C.
Proton charge = $+1.6 \times 10^{-19}$C.
$\varepsilon_0 = 8.85 \times 10^{-12}$ F m^{-1}.

Section 10

Waves and Vibrations (II)

Concept Mapping

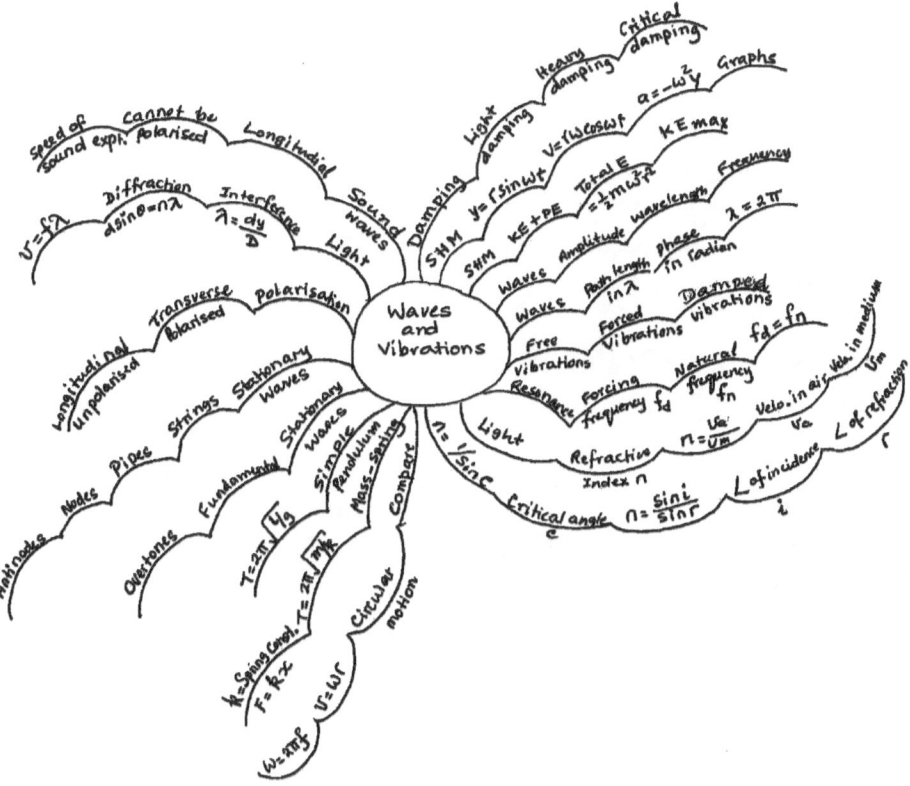

Revisiting Waves

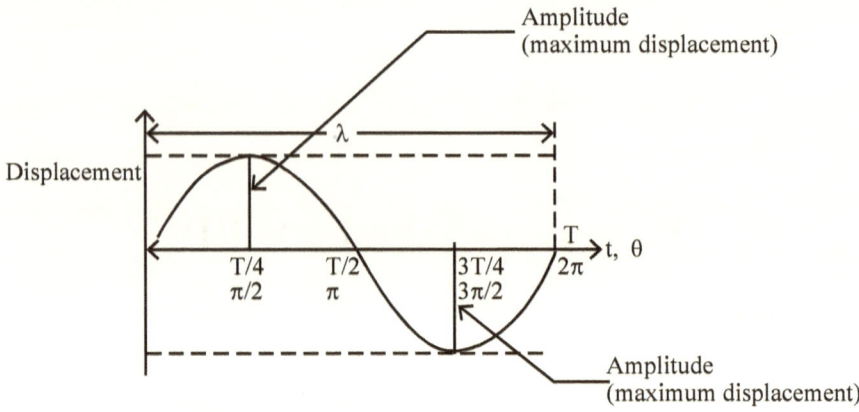

Wave velocity $v = \lambda/T = f\lambda$
   Frequency $f = 1/T$
   Time period $T = 1/f$

From circular motion, angular velocity $\omega = \theta/t$, at any time t
   $\theta = \omega t$.

Linear velocity $v = 2\pi r/T = 2\pi rf$
   $\omega = 2\pi/T = 2\pi f$
   $v = \omega r$.

A wave is a mode of dissipating energy. Wave as in the case of tsunami and seismic vibrations dissipate extremely large amount of energy, resulting in enormous destruction. Waves are produced due to the disturbance of a medium.

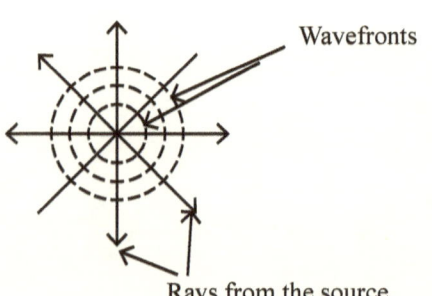

Rays from the source

Dropping a stone into the water in a pond will produce circular waves with the centre where the stone was dropped.

The rays are perpendicular to the wavefronts.

The distance between two adjacent wavefronts gives the wavelength (λ).

There are two types of waves: transverse and longitudinal. All electromagnetic waves are transverse, in which the particles vibrate at 90° to the direction of the wave motion. Sound wave is longitudinal, in which the particles in the medium vibrate parallel to the direction of sound propagation. Sound wave is due to compressions and rarefactions of the particles in the medium.

Transverse Wave

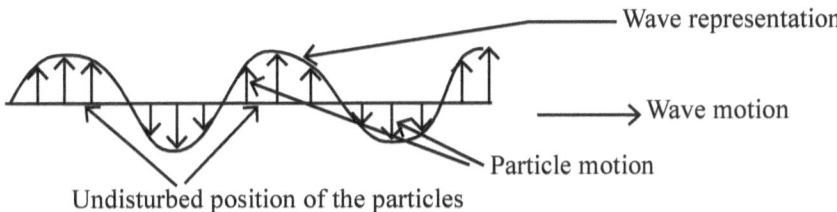

Longitudinal Wave

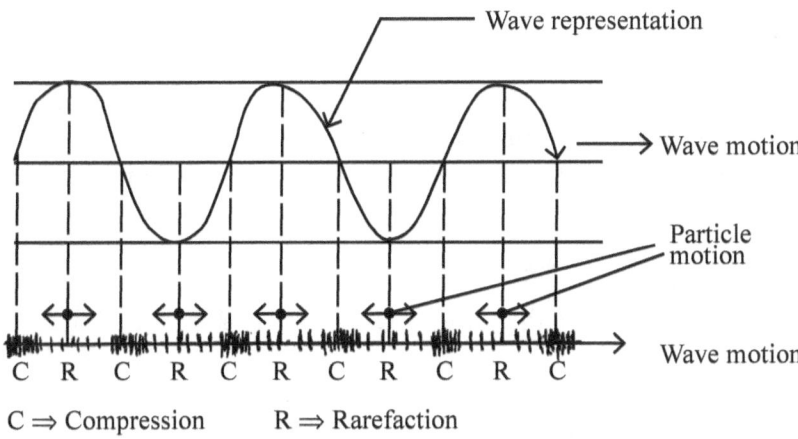

$C \Rightarrow$ Compression $R \Rightarrow$ Rarefaction

Free oscillation $\Rightarrow$ where the amplitude remains constant.
Forced oscillation $\Rightarrow$ where oscillations are caused by an external energy source.

Every object has a natural frequency fn.

For a forced oscillation, if the driving frequency of the external energy source equals the natural frequency of the driven oscillator, resonance occurs. Then the amplitude of vibration becomes maximum.

At resonance $f_d = f_n$.

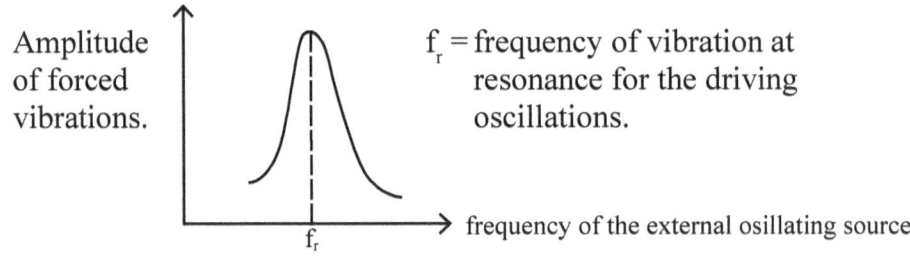

Amplitude of forced vibrations.

f_r = frequency of vibration at resonance for the driving oscillations.

frequency of the external osillating source.

Damping

Damping can be introduced to a mass-spring oscillating system by having a bigger mass at the end of the spring.

When damping is present, the condition of resonance changes. Under damped situation, the amplitude of the forced vibrations decreases. Damping also reduces the natural frequency of the driven oscillations.

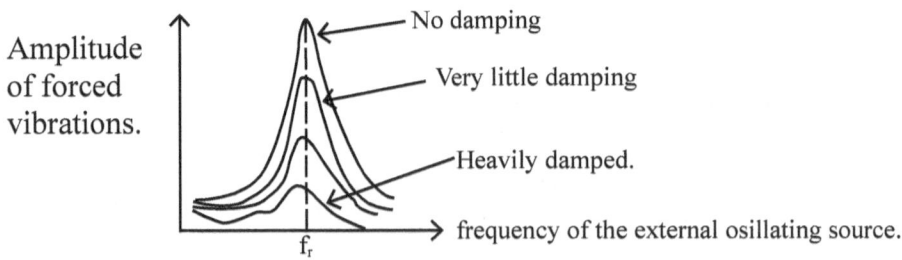

Amplitude of forced vibrations.

No damping

Very little damping

Heavily damped.

frequency of the external osillating source.

For any vibrating system, usually damping is present, such as resistance of the air. The vibrating system loses energy into the surroundings. Damping reduces the amplitude and affects the condition of resonance.

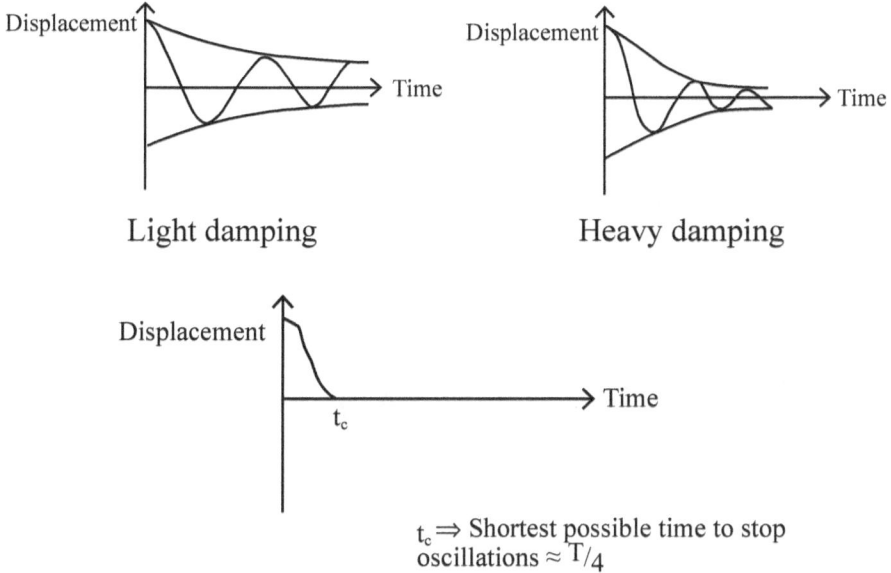

Light damping   Heavy damping

$t_c \Rightarrow$ Shortest possible time to stop oscillations $\approx T/4$

Critical damping

All transverse waves exhibit following properties:

- Reflection
- Refraction
- Diffraction
- Interference
- Polarisation.

Remember, longitudinal waves cannot be polarised.

All electromagnetic waves are transverse. Light is a transverse electromagnetic wave. Maxwell developed the electromagnetic theory and showed that light is made up of electric and magnetic components vibrating at right angles to each other. By using a polarising filter, we can show that vibrations in one particular plane can pass through when we rotate the filter slowly. So, light can be polarised in a plane. Speed of light in free space is given by

$c = \sqrt{1/(\mu_o \varepsilon_o)}$ where μ_o = permeability of free space

ε_o = permittivity of free space

$\mu_o = 4\pi \times 10^{-7}$ Hm^{-1} and $\varepsilon_o = 8.85 \times 10^{-12}$ Fm^{-1}.

Waves carry energy through the medium. The intensity of the wave is the energy flowing per unit area at 90° to the direction of wave motion.

A point wave source emanates energy in all possible directions in space. Therefore, the area at a distance r from the point source will be the surface area of a sphere $4\pi r^2$.

Intensity of a point source $I = P/4\pi r^2$ (Wm^{-2})
$\Rightarrow I \propto 1/r^2 \Rightarrow$ inverse-square law.

Refraction of light

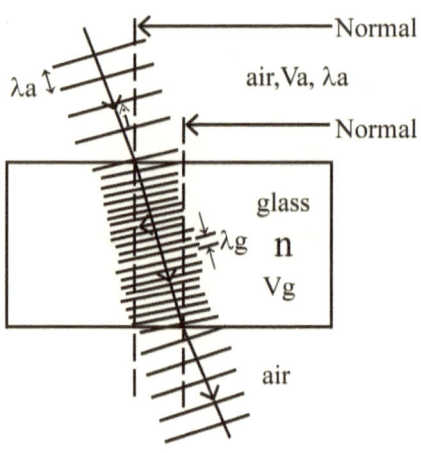

Refractive index of glass
$n_g = \sin i/\sin r$ (i = angle of incidence, r = angle of refraction)

Also, $n_g = v_a/v_g = 3 \times 10^8 \ ms^{-1}/2 \times 10^8 \ ms^{-1}$
$v_a > v_g$
$\lambda_a > \lambda g$
$n_{g = 1.5}$

Critical Angle

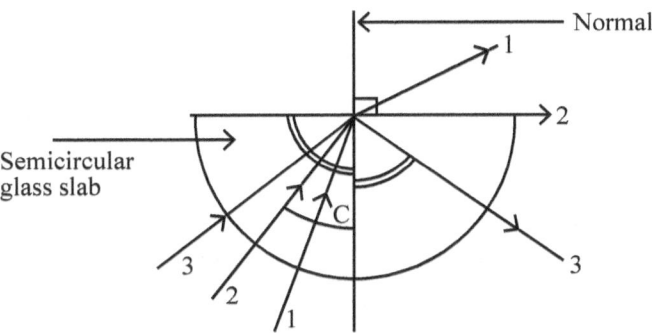

Consider light rays being incident at the centre of the semicircular glass slab, travelling from glass to air.

> For ray 1, $r > i \Rightarrow$ normal refraction.
> For ray 2, $r = 90°$
> and $i = C$ = critical angle for glass.
> n_g = sin (angle in air)/sin (angle in glass) = $\sin 90°/\sin C$
> $\therefore n_g = 1/\sin C$
> $\Rightarrow \sin C = 1/n_g$
> $\Rightarrow \sin C = 2/3 \ (n_g \approx 1.5)$
> $C = 41.8°$ for glass.

For ray 3, there is no refraction. The ray is totally reflected within the glass slab, following laws of reflection. This is called total internal reflection (TIR).

Two conditions must be satisfied to produce TIR.

- The ray must travel from a denser medium (e.g. glass) to a rarer medium (e.g. air).
- The angle of incidence in the denser medium must be bigger than the critical angle of that medium.

An optical fibre has a core of high refractive index material (glass or plastic) surrounded by a cladding of lower refractive index than that of the core. By using the principle of TIR, light signals can travel through the optical fibre long distances and around any bends.

Advantages

Many gigabytes of information can be sent using only a small cable housing a bundle of optical fibres.

There is no interference or noise as in electrical signals. There is hardly any energy loss due to heating effect.

Inter-continental fibre-optic cables are laid at the seabed for effective information exchange.

All fibre-optic cables are underground, not occupying any space in the environment.

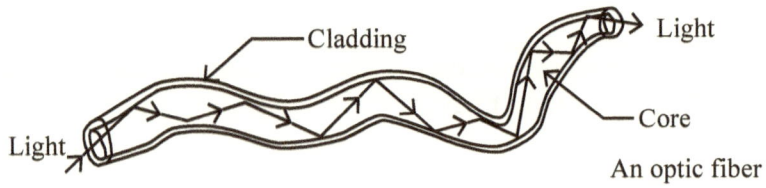

An optic fiber

Diffraction of Light

When a single colour (single λ and hence single f) light is allowed to go through a diffraction grating in which there are thousands of slits per cm, there appear brighter and darker bands on the screen.

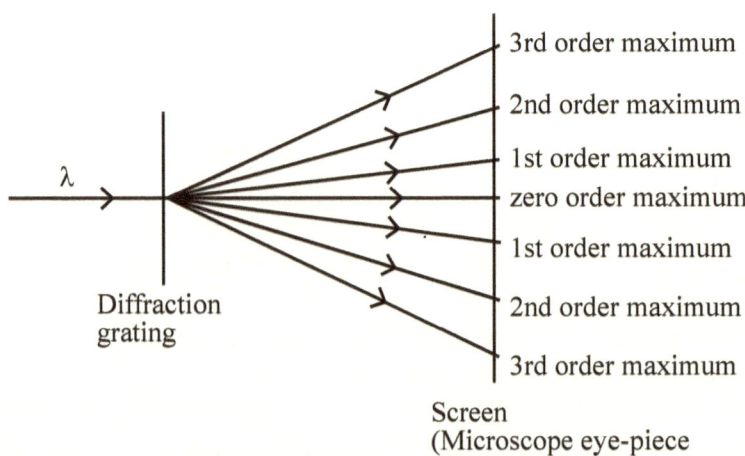

Each slit on the grating produces diffraction.

Each diffraction pattern then superimposes to produce interference.

The overall sharper maximum and minimum pattern is the result of interference.

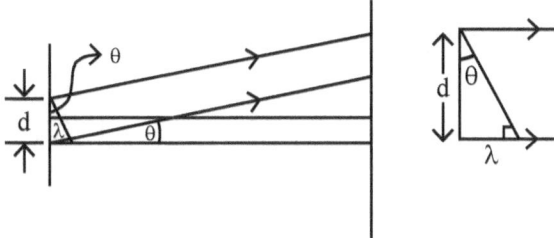

For the first order diffraction maximum, the path difference of diffracted rays from two adjacent slits is λ.

∴ $d \sin\theta = \lambda$, where d = slit separation
θ = angle of diffraction
$d = 1/N$ (N = no. of lines on the grating).

The general diffraction equation is $d \sin\theta = n\lambda$ (n = diffraction order).

For bigger λ, $\sin\theta$ is bigger, θ is bigger.
For bigger d, $\sin\theta$ is smaller, θ is smaller.

Stationary Waves

When a wave in motion meets a boundary, it gets reflected.

If the time for the wave to travel back and forth is same, then incoming and reflected waves superimpose and the stationary wave is produced. The frequency at which stationary waves are produced is the resonant frequency.

Stationary waves can be produced in a vibrating string and vibrating air columns in pipes.

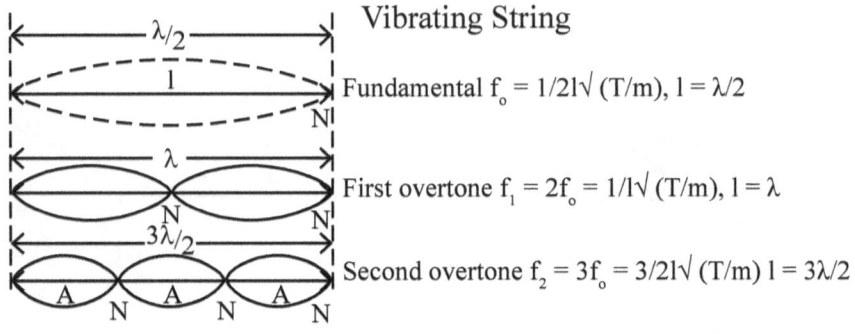

Vibrating String

Fundamental $f_0 = 1/2l\sqrt{(T/m)}$, $l = \lambda/2$

First overtone $f_1 = 2f_0 = 1/l\sqrt{(T/m)}$, $l = \lambda$

Second overtone $f_2 = 3f_0 = 3/2l\sqrt{(T/m)}$ $l = 3\lambda/2$

N = nodes   l = length of the string
A = antinodes   T = tension on the string
                           m = mass of the sting per m.

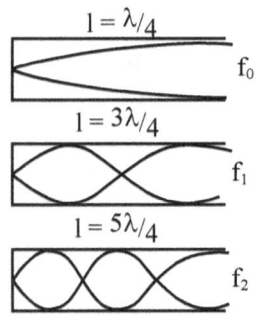

Pipe open at one end,
Nodes at closed end,
Antinodes at open end.

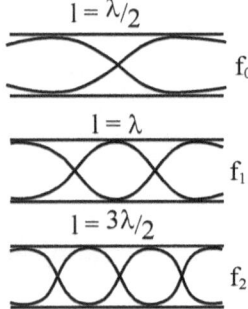

Pipe open at both ends,
Antinodes at open ends,
Nodes inside the pipe.

Simple Harmonic Motion

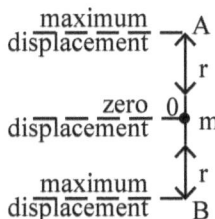

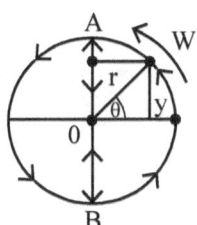

Vertical oscillations of a small mass attached to a spring.

Circular motion of mass m at the end of a rotating vector r is equivalent to the vertical oscillations of mass m at the centre of the circle.

Displacement $y = r \sin\theta$
$$y = r \sin\omega t \ (\theta = \omega t).$$
Velocity $v = dy/dt = r\omega\cos\omega t.$
Acceleration $a = dv/dt = -r\omega^2 \sin\omega t$
$$a = -\omega^2 y.$$

Therefore, acceleration is directly proportional to the displacement and is always directed toward the centre.

This is the definition of a system undergoing simple harmonic motion.

Maximum displacement $y_{max} = r$ at A and B
Zero displacement at the centre O.
Maximum velocity $v_{max} = \omega r$ at centre O
Zero velocity at A and B.
Maximum acceleration $a_{max} = |\omega^2 r|$ at centre O
Zero acceleration at A and B.

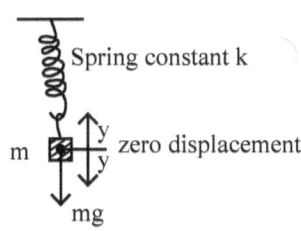

Spring constant k

zero displacement

m

mg

Applying Hooke's law to the mass-spring oscillator, extension produced is directly proportional to the load applied.

$$mg \propto y$$
$$mg = ky, \; k = \text{spring constant} = mg/y \; (N/m)$$
$$g = (k/m) \, y.$$

Compare $a = |\omega^2 y|$

$$\therefore \omega^2 = k/m$$
$$\Rightarrow 4\pi^2/T^2 = k/m$$
$$\Rightarrow T^2 = 4\pi^2 m/k.$$

T depends on k and m only. $T^2 \propto m$, and $T^2 \propto 1/k$.

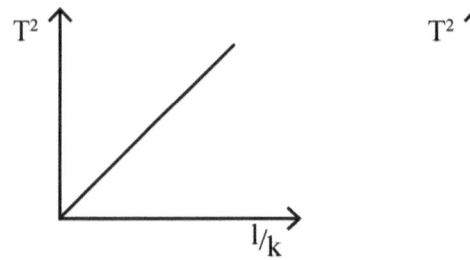
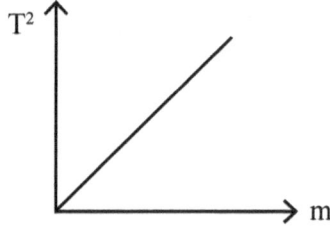

A simple pendulum is a simple harmonic oscillator.

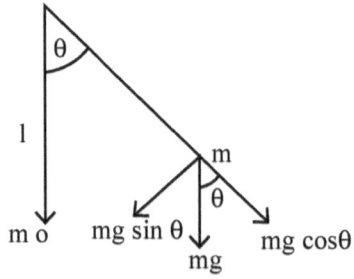

The pendulum bob can be thought of moving in a circle of radius r with an angular velocity ω.

The component $mg \sin\theta$ is the centripetal force. $\sin\theta \approx \theta$ when θ is a very small angle in radians.

$$mg\theta = m\omega^2 r$$
$$g\theta = \omega^2 r = \omega^2 l\theta \; (r = l\theta)$$
$$g = 4\pi^2 l/T^2$$
$$T^2 = 4\pi^2 \, l/g$$
$$T = 2\pi\sqrt{(l/g)}$$
$$T^2 \propto l \text{ and } T^2 \propto l/g.$$

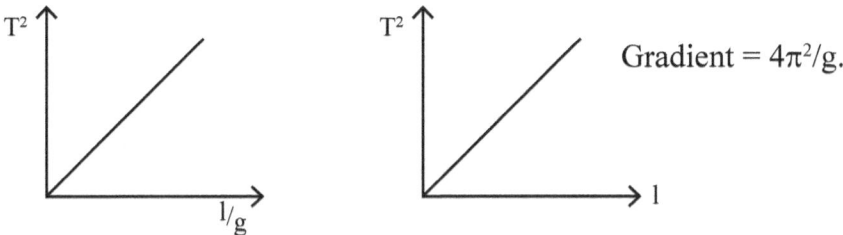

Gradient = $4\pi^2/g$.

From a series of values for T^2 and l, the gradient of the second graph is measured. A value for g can be calculated from the gradient.

SHM graphs

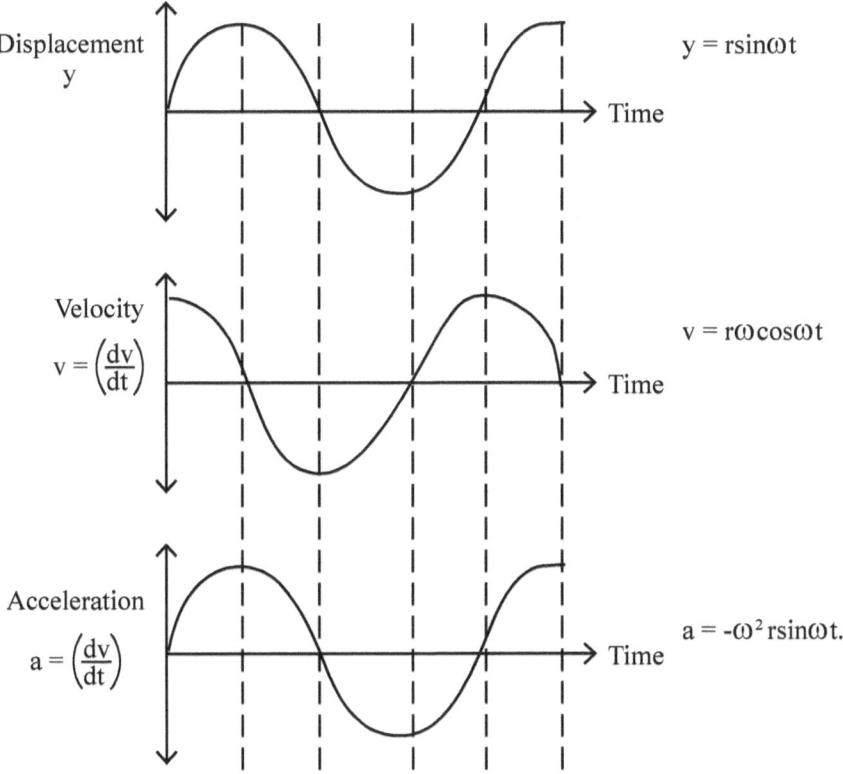

$y = r\sin\omega t$

$v = r\omega\cos\omega t$

$a = -\omega^2 r\sin\omega t.$

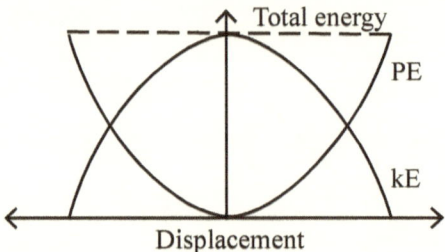

Energy exchange of the oscillator.

Total energy

Energy exchange of a simple harmonic oscillator.

Instantaneous PE = $\frac{1}{2}$ ky^2 (k = spring constant)
PE$_{max}$ = $\frac{1}{2}$ kr^2. (r = amplitude)
Instantaneous KE = $\frac{1}{2}$ m(dy/dt)^2
KE$_{max}$ = 1/2m v$^2_{max}$
KE$_{max}$ = $\frac{1}{2}$ m ω^2r^2.

At any point of the oscillator, total energy remains constant.
Total energy = KE$_{max}$ = $\frac{1}{2}$ mω^2r^2.

Learning Zone

Progressive waves carry energy.
Intensity is the amount of energy flowing per second per unit area.
Intensity $\propto$ (amplitude)^2.
Path difference λ = phase difference 2π.
Damping of an oscillator is due to energy loss to the surroundings.
Damping reduces amplitude and resonant frequency.
Refractive index n = sini/sinr
                   n = 1/sinC where C is the critical angle.
Resonance condition$\Rightarrow$f$_d$ = f$_n$.
Stationary waves are caused by reflected waves falling on incoming waves.
Stationary waves produce nodes and antinodes.
Nodes $\rightarrow$ points of no vibration.
Antinodes $\rightarrow$ points of maximum vibration.
For a mass-spring oscillator   T = $2\pi\sqrt{}$ (m/k) T = time period, m = mass on the spring.
For a simple pendulum   T = $2\pi\sqrt{}$ (l/g) T = time period.

For a vibrating string $f = n/2l\sqrt{(T/m)}$ T = tension and
$n = 1,2,3$, etc.
m = mass/length.

Vertical displacement of SHM $y = r\sin\omega t$

Velocity $v = dy/dt = r\omega\cos\omega t$

Acceleration $a = dv/dt = -\omega^2 r\sin\omega t$.

$PE_m = \frac{1}{2} kr^2$

$KE_{max} = \frac{1}{2} m\omega^2 r^2$.

For diffraction maxima $d\sin\theta = n\lambda$ ($n = 1,2,3$, etc.)

Section 11

Mechanical Properties of Materials

Concept Mapping

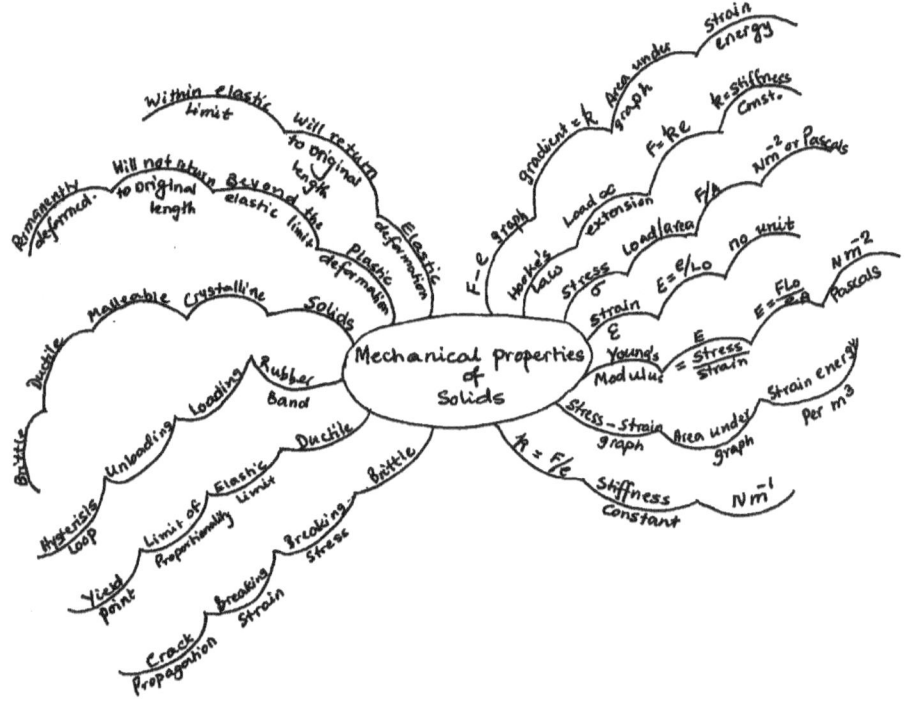

Hooke's law $\Rightarrow$ Load (force) applied to stretch the material is directly proportional to the extension produced.

$F \propto e \Rightarrow F = ke$
$\Rightarrow k = F/e \ (\text{Nm}^{-1})$.

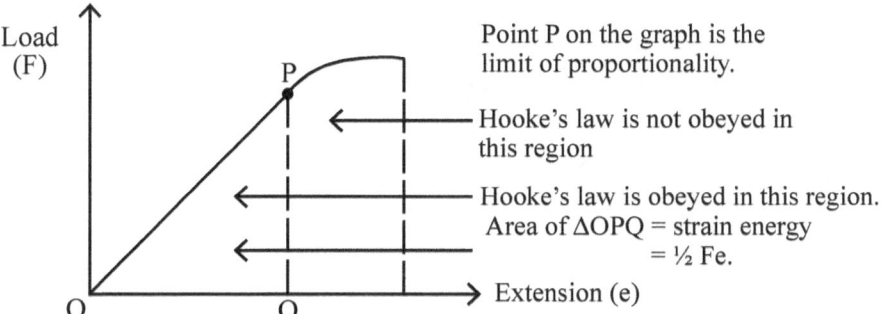

Point P on the graph is the limit of proportionality.

Hooke's law is not obeyed in this region

Hooke's law is obeyed in this region.
Area of $\triangle OPQ$ = strain energy
= $\frac{1}{2}$ Fe.

Strain energy = $\frac{1}{2}$ Fe
= $\frac{1}{2}$ ke².

This is the elastic potential energy stored in the stretched sample.
Stress = load/area = F/A (Nm⁻²).
Strain = extension/original length = e/L_o
Young's modulus E = stress/strain
E = FL_o/Ae (Nm⁻²) or pascals (Pa)

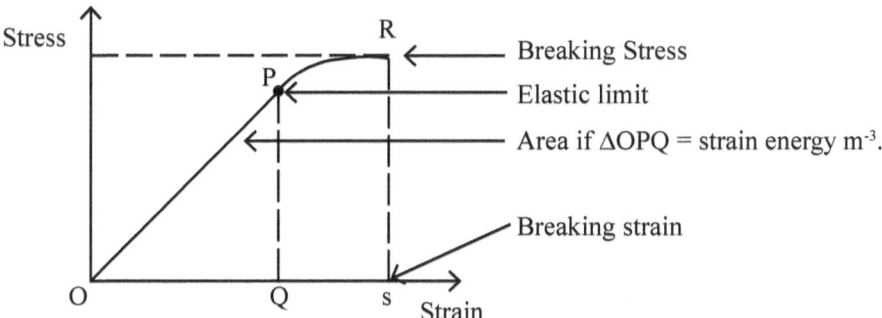

Breaking Stress

Elastic limit

Area if $\triangle OPQ$ = strain energy m⁻³.

Breaking strain

Area of $\triangle OPQ$ = $\frac{1}{2}$ × stress × strain
= $\frac{1}{2}$ × F/A × e/L_o
= $(1/2\ Fe)/AL_o$
= strain energy/volume.

Gradient of the lien OP = stress/strain
= Young's modulus (E).

Stress-strain graph for a copper wire (ductile material).

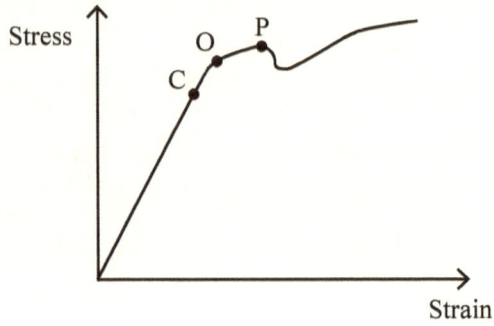

Point C⇒ Limit of proportionality. Hooke's law is obeyed up to this point.

Point O ⇒ Elastic limit. Beyond this point plastic deformation occurs.

Point P ⇒ Yield point. The material extends without any extra load beyond that point.

Stress-strain graph for a glass fibre (brittle material).

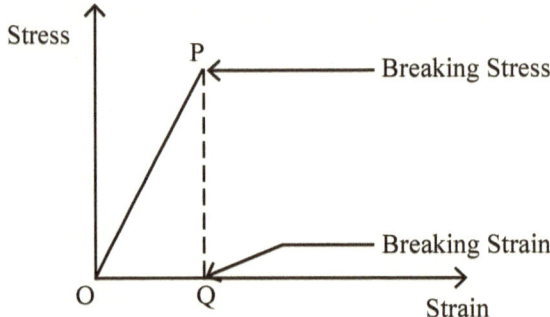

At point P, the material snaps. This is called brittle fracture. The material obeys Hooke's law. There is no plastic deformation beyond point P. Under tension, cracks in brittle materials grow.

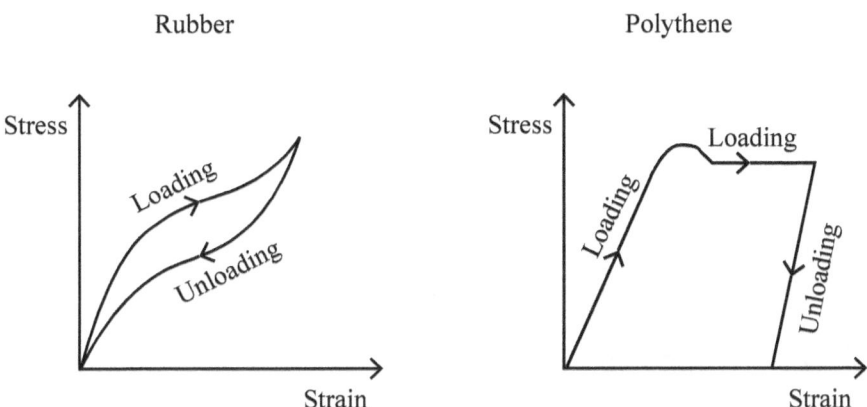

Under loading and unloading, rubber behaves elastically in that it returns to its original length when the load is removed. Under same operations, polythene behaves plastically, in that it does not go back to its original length when the applied load is removed. The area within the loops of each graph is the amount of work done on the sample during the loading-unloading process. This is the energy stored in the sample.

Section 12

Electrical Energy

Concept Mapping

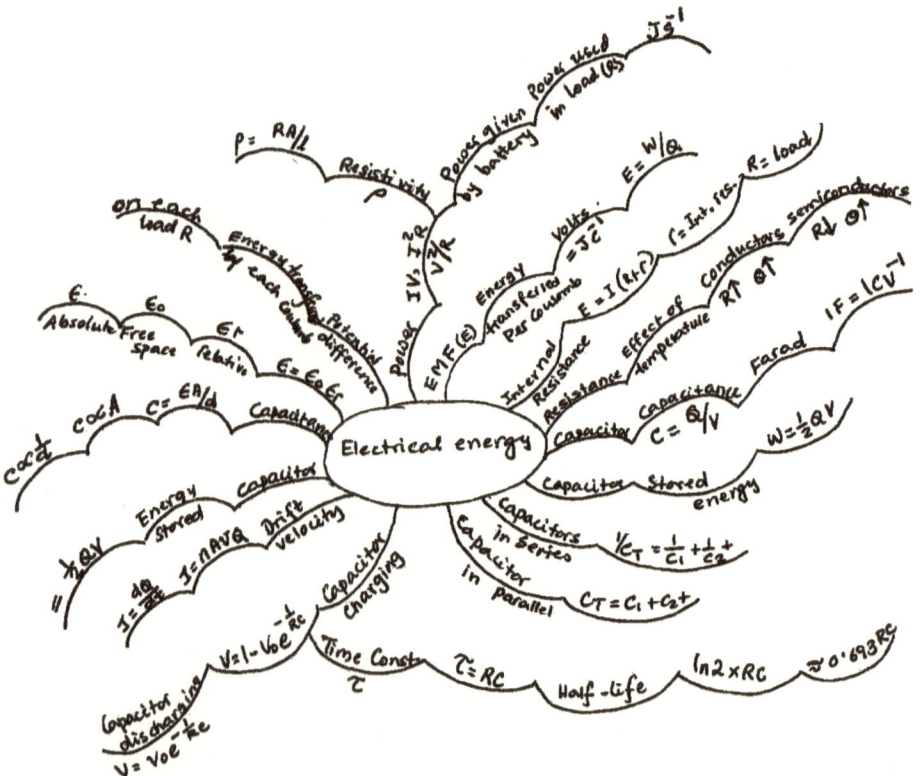

Electrical energy

emf $\Rightarrow$ electromotive force = energy transferred per coulomb in the battery.

Potential difference = energy transferred per coulomb in the load (R).

1 volt = $1JC^{-1}$
1 amp = $1CS^{-1}$
1 watt = $1JS^{-1}$
Ohm's law $\Rightarrow$ V = RI when the temperature of the component is
constant.

Power = VI = I^2R = V^2/R in watts or JS^{-1}.
Energy = power × time = VIt in joules.

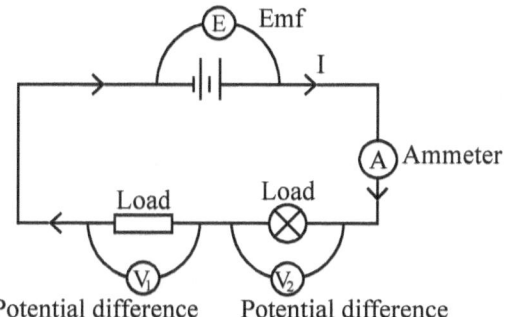

Potential difference   Potential difference

Without any internal
resistance of the battery,

$$E = V_1 + V_2$$

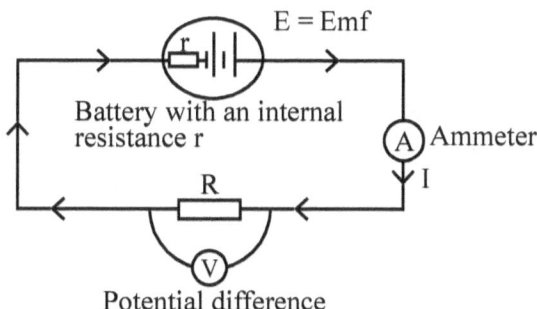

Potential difference

Internal resistance is caused by the chemical reaction in the battery when
the current is set up in the circuit.

E = IR + Ir
E = V + Ir   E = emf of the battery, V = potential difference
V = E − Ir

By changing the value of load resistance R, a series of values for V and
I can be obtained to plot a graph.

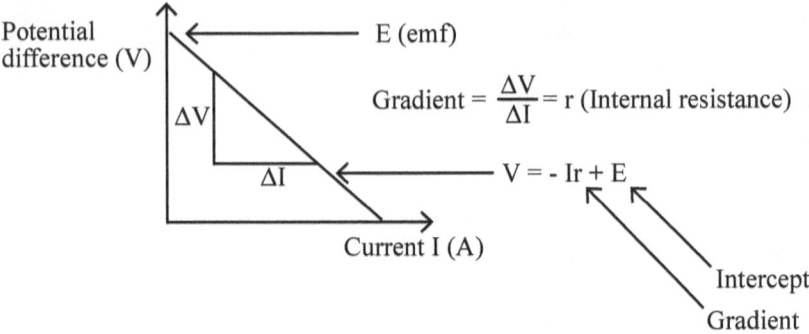

Gradient = V/I = r (internal resistance)

The current in a circuit is caused by the flow of charge carried by the electrons in the conducting wire.

Current is number of charges flowing per second. $I = dq/dt$
One electron carries a charge of 1.6×10^{-19} C.

The flow of electrons is restricted by the vibrations of the atoms in the conducting wire. The overall velocity of the electrons is called drift velocity V.

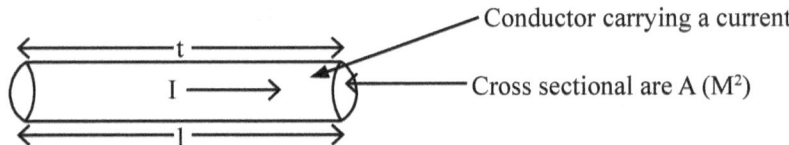

n = no. of electrons per m^3
q = charge on each electron
Volume = Al (m^3)
Drift velocity = l/t (ms^{-1})
I = Total charge/time = nAlQ/t
∴ I = nAVQ.

$R \propto l$ and $R \propto 1/A \Rightarrow R = \rho\, l/A$
ρ = resistivity of the material, the unit is ohm-metre (Ωm)
ρ = RA/l, conductivity $\sigma = 1/\rho = 1/RA$, the unit is $(\Omega m)^{-1}$
ρ depends on resistance and hence on light intensity and temperature.

At higher temperatures, atomic vibrations become more and the resistance to charge flow increases. Power loss as a result of heating is I^2R.

At low temperatures, atomic vibrations decrease, lowering the resistance to charge flow. If the temperature is lowered below a temperature called 'transition temperature' for the material, it becomes a superconductor, without any resistance. Where a large current is required, conducting wires are cooled down to liquid helium temperature (about −269°C) to allow a very large current to flow.

Current–voltage Graphs Revisited

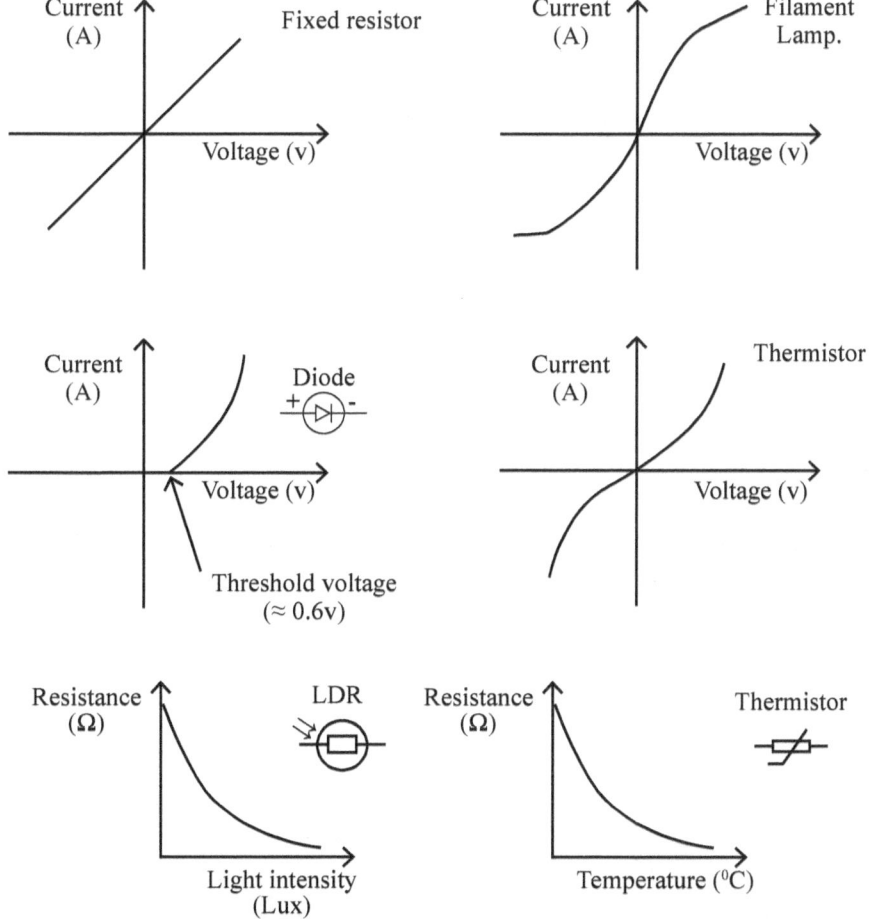

A Fixed Potential Divider

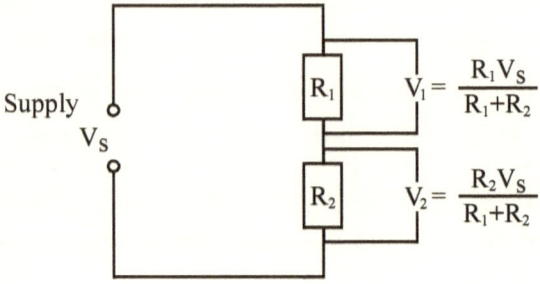

Current $I = Vs/(R_1 + R_2)$ (Vs = supply voltage)

Potential difference $V_1 = R_1 I$

Potential difference $V_2 = R_2 I$

V_1 and V_2 depend on the chosen values of R_1 and R_2 for a fixed supply voltage Vs.

A Continuous Output Potential Divider

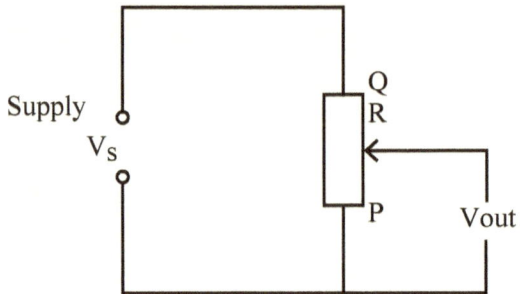

The position of the slider at R gives the Vout as the PD across RP.

   At P, Vout = 0

   At Q, Vout = Vs

Capacitors

A capacitor is a charge storing device. Two conducting plates connected to positive and negative terminals of a battery with a gap between the plates make a simple capacitor. Any material between the two plates is called a dielectric (e.g. plastic, air, mica). Capacitors are used in timing circuits.

Capacitance C of a capacitor is the amount of charge stored Q divided by the voltage V.

$$C = Q/V \text{ (farad F)}$$
$$1F = ICV^{-1}.$$

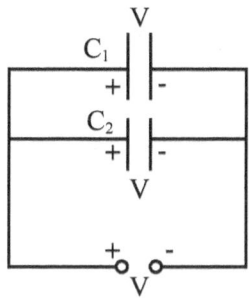

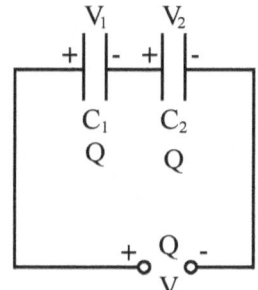

Two Capacitors in Parallel

$$Q_{total} = Q_1 + Q_2$$
$$\Rightarrow C_{total} V = C_1 V + C_2 V$$
$$\therefore C_{total} = C_1 + C_2.$$

Two Capacitors in Series

Charge flowing in the circuit is constant, but voltage across each capacitor is different.

$$V = V_1 + V_2$$
$$\Rightarrow Q/C_{total} = Q/C_1 + Q/C_2$$
$$\therefore 1/C_{total} = 1/C_1 + 1/C_2.$$

1 farad is a big capacitance.
Values used for practical capacitance are
  $1\mu F = 1\text{microfarad} = 10^{-6}F$
  $1nF = 1\text{nanofarad} = 10^{-9}F$
  $1pF = 1\text{picofarad} = 10^{-12}F.$

Capacitance C ∝ area of the plates
          $C \propto A$
Capacitance C ∝ 1/d (d = plate separation)
Capacitance C ∝ ε (ε = absolute permittivity of the dielectric medium between the plates)
  $\varepsilon = \varepsilon_0 \varepsilon_r$ where ε_0 = permittivity of free space = 8.85×10^{-12} Fm^{-1}
  ε_r = relative permittivity
  $C = \varepsilon A/d = \varepsilon_0 \varepsilon_r A/d.$

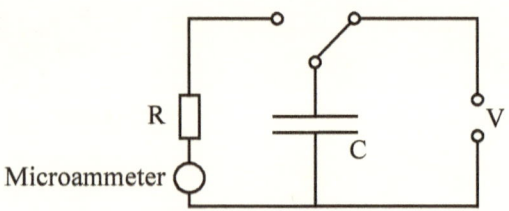

Discharging a Fully Charged Capacitor Through a Resistor

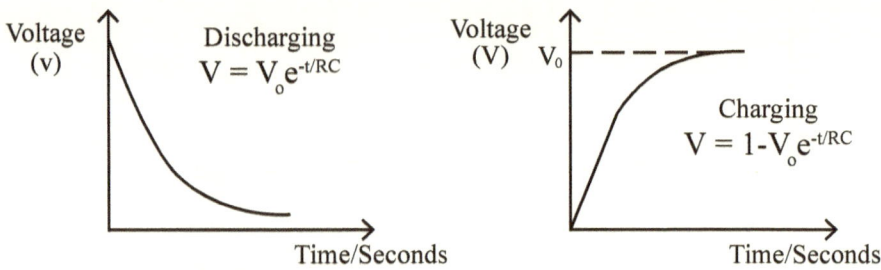

Stored charge in the capacitor decays exponentially. This means the time taken to go to half the value of charge at any instant is constant.

$Q = Q_0 e^{-t/RC}$
Q = charge remaining at any time t
Q_0 = charge at t = o
RC = Time constant τ.
When $t = RC \Rightarrow Q = Q_0 e^{-1}$
$Q/Q_0 = 1/e = 1/2.718 = 0.37 \Rightarrow Q = 0.37 Q_0$.

Therefore the time constant is the time taken for the charge to fall to 37% of original charge Q_0.

For a charging capacitor this is also the time for the charge to rise to 63% of original charge Q_0.

The time taken for a capacitor to fully charge or fully discharge is accepted to be about 5RC.

To find half-life, put $t = t_{1/2}$ and $Q_0/Q = 2$

$Q_0/Q = e^{t/RC}$

$\Rightarrow 2 = e^{t_{1/2}/RC}$

$\Rightarrow l_n 2 = t_{1/2}/RC$

$\therefore t_{1/2} = l_n 2 \times RC$

$t_{1/2} = 0.693\ RC.$

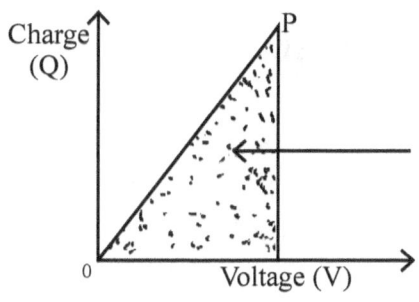

Gradient = capacitance of the capacitor.

Area of $\triangle OPN$ = energy stored in the capacitor

$= \tfrac{1}{2}QV.$

The energy supplied by the battery is QV. Half of this energy is lost as heat in the conducting wire of the circuit.

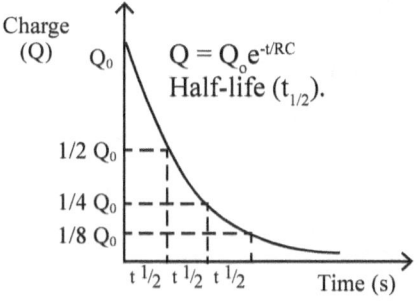

$Q = Q_0 e^{-t/RC}$
Half-life ($t_{1/2}$).

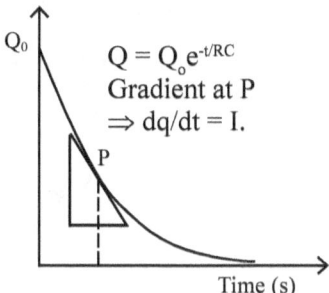

$Q = Q_0 e^{-t/RC}$
Gradient at P
$\Rightarrow dq/dt = I.$

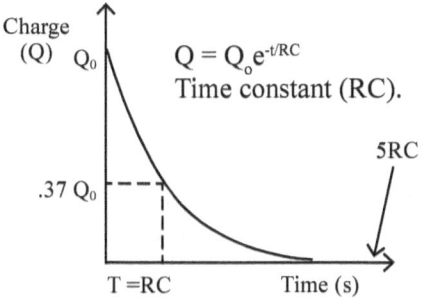

$Q = Q_0 e^{-t/RC}$
Time constant (RC).

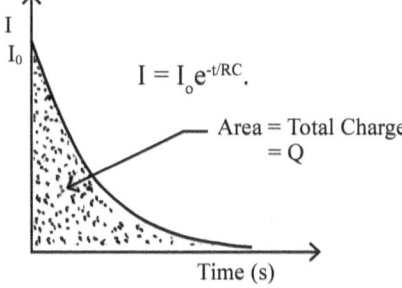

$I = I_0 e^{-t/RC}.$

Area = Total Charge
$= Q$

Section 13

Atoms and Quanta

Concept Mapping

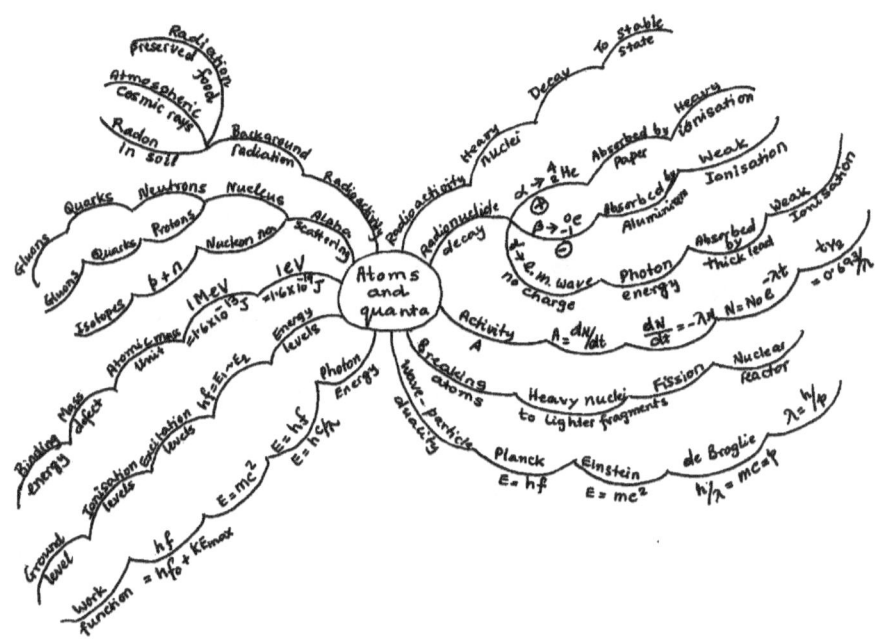

Radioactivity

Activity = dN/dt (rate of decay = number decaying per second)

dN/dt is directly proportional to the number of radionuclide present

$\Rightarrow$ dN/dt $\propto$ N

$\Rightarrow$ dN/dt $= -\lambda$N (negative sign signifies decay)

$\Rightarrow$ (1/N) dN $= \lambda$dt

Integrating both sides with the limits, N: N_o to N and t: 0 to t

$\Rightarrow l_n N - l_n N_o = -\lambda t$

$$\Rightarrow l_n (N/N_o) = -\lambda t$$
$$\Rightarrow N/N_o = e^{-\lambda t}$$
$$\therefore N = N_o e^{-\lambda t}$$
$$\therefore A = A_o e^{-\lambda t} \ (A \propto N).$$

Activity is defined as disintegration per second and the unit is Bq. 1Bq = 1 disintegration per second.

Half-life of a radioactive nucleus is the time taken for its activity to become half.

At $t = t_{1/2}$, $N_o/N = 2$. or $A_o/A = 2$.
$$A_o/A = e^{\lambda t}$$
at $t = t_{1/2}$ $2 = e^{\lambda t 1/2}$
$$\Rightarrow l_n 2 = \lambda t_{1/2}$$
$$\Rightarrow t_{1/2} = l_n 2/\lambda$$
$$\Rightarrow t_{1/2} = 0.693/\lambda.$$

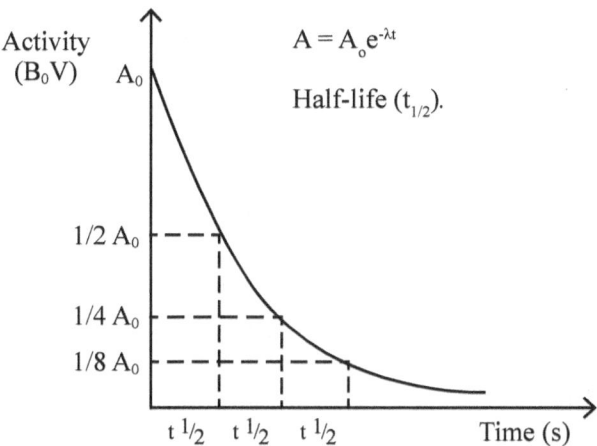

After 1 half-life, activity becomes $1/2 = 1/2^1$
After 2 half-lives, activity becomes $1/4 = 1/2^2$
After 3 half-lives, activity becomes $1/8 = 1/2^3$
After n half-lives, activity becomes $= 1/2^n$

A nucleus is formed from its constituents called nucleons. The energy required in MeV (one million electron-volt $= 1.6 \times 10^{-13}$ J) to separate all the nucleons in a nucleus is called the binding energy. This is equal to the mass defect, which is the energy released in forming the nucleus.

Binding energy ≡ mass defect

1 atomic mass unit = 1.66×10^{-27} kg
1 eV = 1.6×10^{-19} J
1 MeV = 1.6×10^{-13} J
Velocity of light c = 3×10^8 ms^{-1}.
Einstein energy E = mc^2.

Binding energy per nucleon $= \dfrac{\text{binding energy}}{\text{nucleon no}} = \dfrac{\text{binding energy}}{\text{mass no.(A)}}$

Fission

$235U92 + 1n0 \rightarrow 144Ba56 + 90Kr60 + 2(1n0) +$ Energy (E)
E = 8×10^{13} J per kg of uranium
E = 200 MeV for one uranium atom.

Fission reaction occurs in a nuclear reactor. Spontaneous radioactive decay also occurs for unstable heavy nuclei.

Fusion

$2H1 + 2H1 \rightarrow 3He2 + 1n0 +$ Energy (E)

E = 3.27 MeV for these two deuterium nuclei.

In fusion reaction, a great amount of energy is required to fuse two smaller nuclei to form a larger nucleus. Nuclei are positively charged, and hence they have coulomb force of repulsion between them. The smaller nuclei have to be brought closer together, overcoming the repulsive force in order to allow strong nuclear interaction to take place so they are fused together. Energy of 1.6×10^{-13} J is needed to overcome the electrostatic repulsion. This can be achieved only at an extremely high temperature, about 8×10^9 K. This is calculated assuming the nuclei to behave as an ideal gas using the formula for KE,

KE = (3/2) kT
KE = 1.6×10^{-13} J.
k = Boltzmann constant = 1.38×10^{-23} JK^{-1}
T = absolute temperature.

Binding energy per nucleon (MeV)

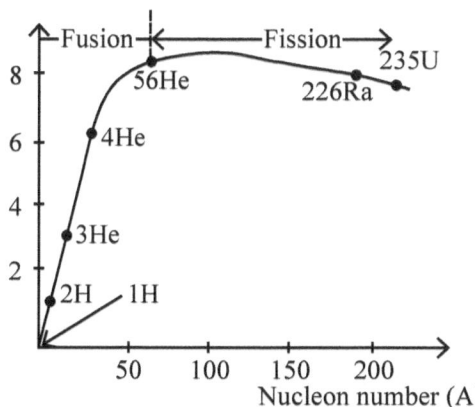

BE – Nucleon number graph

Up to 56 Fe, most of the nuclei are stable. It is the fusion region. Above 56 Fe majority nuclei are unstable. It is the region of fission reaction.

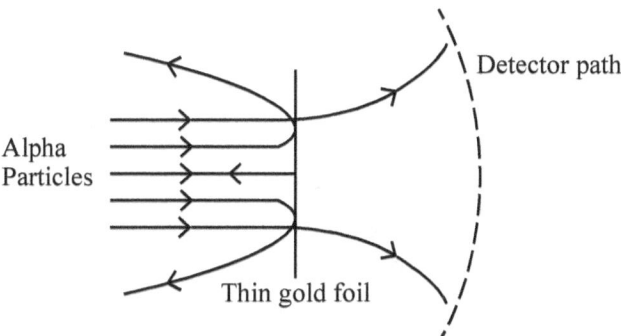

Rutherford's α-particle scattering

By bombarding a thin gold foil by α-particles, Rutherford found that some α-particles bounce back from the foil at 180°. This shows the presence of a positively charged nucleus in the atom.

Photoelectric Emission

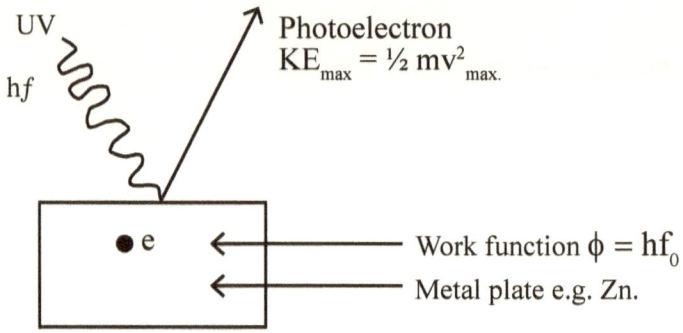

There is a minimum energy required to move the electrons in the metal plate to the surface. The frequency corresponding to this energy is called the threshold frequency f_0 below which photoelectric emission does not occur. When the metal plate is hit with a photon energy hf, which is bigger than hf_0, photoelectrons will be emitted with a maximum KE of $\frac{1}{2} mv^2_{max}$.

$\phi = hf_0$ = work function of the metal = minimum energy needed for the photoelectric emission to take place.
hf = photon energy

Einstein's photoelectric equation $\Rightarrow hf = hf_0 + \frac{1}{2} mv^2_{max}$.
The emission of the photoelectrons can be stopped by applying an opposing potential called stopping potential Vs to the flow of photoelectrons.

Then $\frac{1}{2} mv^2_{max} = eVs$ and $eVs = hf - hf_0$
$\Rightarrow Vs = (h/e) f - (h/e) f_0$
$\Rightarrow Vs = hf/e - hf_0/e$
$\Rightarrow Vs = hf/e - \phi/e$.

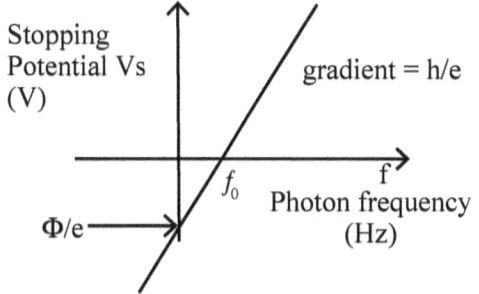

h = Planck's constant
= 6.6×10^{-34} JS

Section 14

Magnetic Fields (II)

Concept Mapping

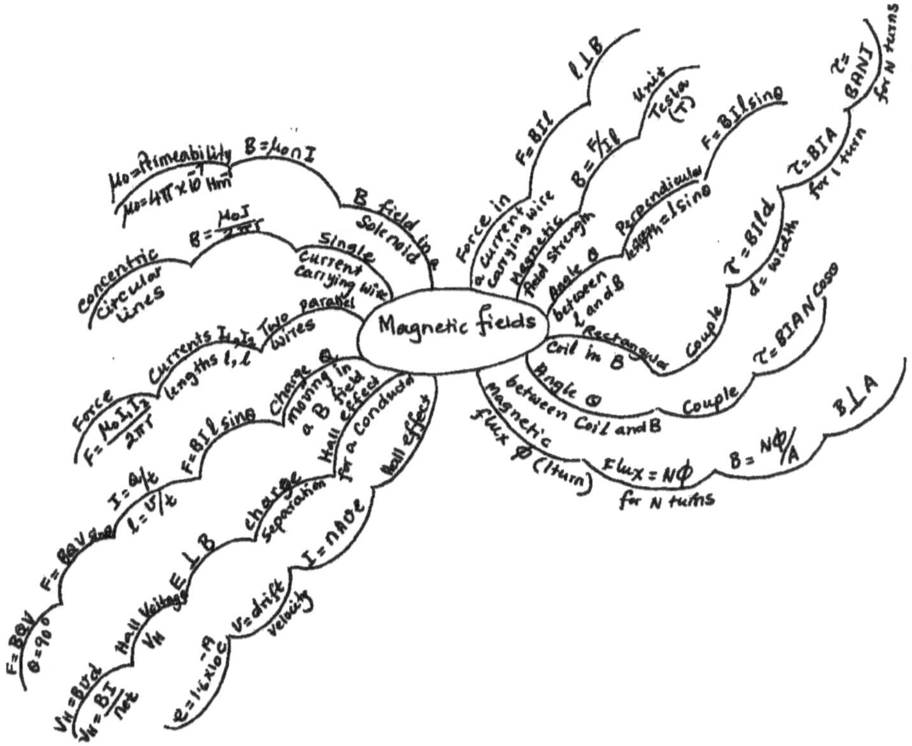

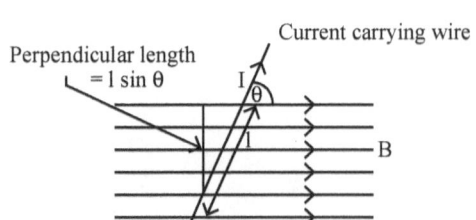

Perpendicular length
= l sin θ

Current carrying wire

B

Force = BIl sinθ in general
F = BIl for θ = 90°
Magnetic field strength
B = F/I l (Tesla T).

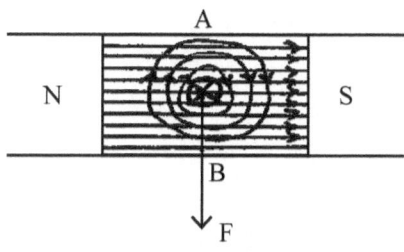

Current in the wire is into the paper $\Rightarrow \otimes$. At A, both the magnetic fields superimpose, making the field stronger. At B, the two magnetic fields cancel each other to produce a weaker field. The conductor moves downwards from strong to weak magnetic field.

F, I, and B are at 90° to each other. The direction of F is given by Fleming's left hand rule.

Transformers

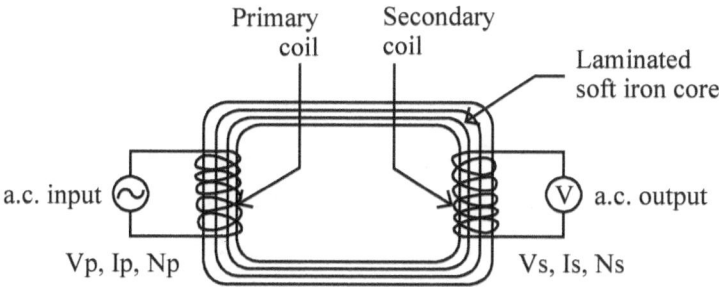

Formulae:
  $Vp/Vs = Is/Ip$.
  $Vp/Vs = Np/Ns$.
  $Is/Ip = Np/Ns$.

Use: A transformer changes working voltages to transmit certain power. Transformers are used in power transmission.

Information:
  • Transformers can work with AC only which produces $d\phi/dt$ (rate of change of flux) to produce induced emf in the secondary coil.

- Transformers work to produce a high voltage and a low current in transmission lines for long-distance transmission to minimise power loss (I^2R) in the cable.
- Transformers are constructed to minimise power loss.

Power loss in a transformer

Due to	How produced?	Where?	How minimised?
Eddy current loss	Due to the changing flux, current loops are produced in a metal	In the soft-iron core	The core is made up of sheets of soft iron joined by insulating glue to minimise current loops produced.
Copper loss	Heat generated in the coils due to the current flow	In the coils of the transformer	By using bigger diameter coils
Flux loss	Not all the flux lines in the primary link the secondary coil.	Flux in the primary linking the secondary	By making the primary and secondary coils closer together with an appropriate soft-iron core
Hysterisis loss	Oscillating AC current produces magnetisation and demagnetisation in the core, which produces heat loss.	In the soft-iron core	By using a core material with a hysterisis loop of smaller area

Movement of a Current-carrying Coil

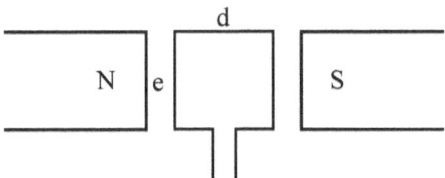

A rectangular coil carrying a current in a B-field rotates due to the force produced in the coil. The force is produced only on the lengths l of the coil, as the widths d are parallel to the B-field.

$F = BIl$

Torque produced $\tau = BIld$ (d = width of the coil, l = length of the coil)
$\Rightarrow \tau = BIA$.

When the coil rotates due to the torque, making an angle θ with B,
$\tau = BIA \cos\theta$ for a single turn of the coil.
$\tau = BIAN \cos\theta$ for N turns.
$\tau = BIAN \cos\omega t$ ($\theta = \omega t$).

Magnetic field strength $B = \Phi/A \Rightarrow \Phi = BA$ (Φ = magnetic flux).
Magnetic field strength B due to a straight wire carrying a current at a perpendicular distance r from the wire $= \mu_o I/2\pi r$.

B due to a solenoid of n turns $= \mu_o nI$.

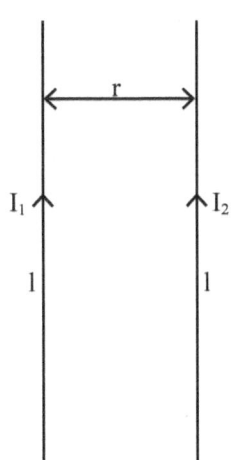

Force produced between two wires of length l separated by a distance r carrying a current I_1 and I_2 respectively.
B due to $I_1 = \mu_o I_1/2\pi r$
Force in $I_2 = BI_2 l$
$F = \mu_o I_1 I_2 l/2\pi r$.

Remember that the B-lines in a current-carrying wire are circular in a plane at right angles to the wire. The magnetic field is stronger nearer the current, and it gets weaker as the distance increases from the current.

$B \propto 1/r \Rightarrow$ inverse proportionality.

The direction of the B-lines is given by right hand grip rule. Point thumb along the current direction; fingers in the grip will give the direction of the B-field.

Hall voltage V_H

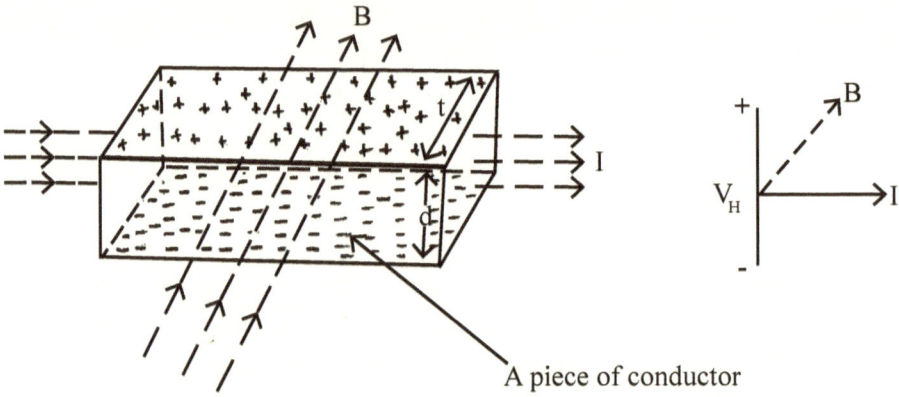

A piece of conductor

Face area of the current flow = d t.

I → left face to right face
B → front face to back face
V_H → top face to bottom face
V_H develops due to the charge separation in the sample.
Electric force = EQ = (V_H/d) Q where E = electric field strength = V_H/d.
Magnetic force = BQv.
In the steady state, electric force = magnetic force
$\Rightarrow (V_H/d) Q = BQv$
$\Rightarrow V_H = Bvd$

I = nAve where v = drift velocity of the electrons
$\Rightarrow v = I/ (nAe) = I/ (ndte)$
$\therefore V_H = BI/(net)$, where n = number of electrons per m^3
e = electronic charge.

If all other variables remain constant, $V_H \propto 1/n$.

So a semiconductor slice where the charge carriers are small in number will produce a bigger V_H.

AC Generator

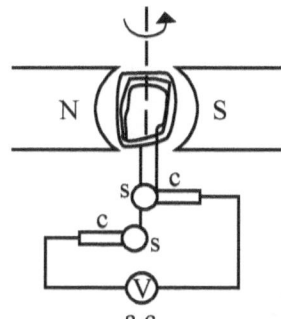

Rotating rectangular coil in a magnetic field.
SS $\Rightarrow$ Slip rings
CC $\Rightarrow$ Carbon brushes
Slip rings stay with the same carbon brushes. This allows current to pass in both directions as the coil rotates with a certain frequency. The output is an alternating voltage.

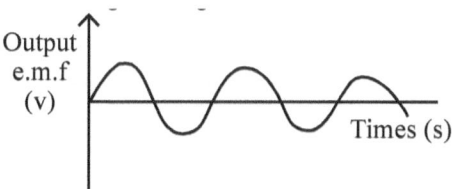

Changing AC to DC.   Half-wave rectification.

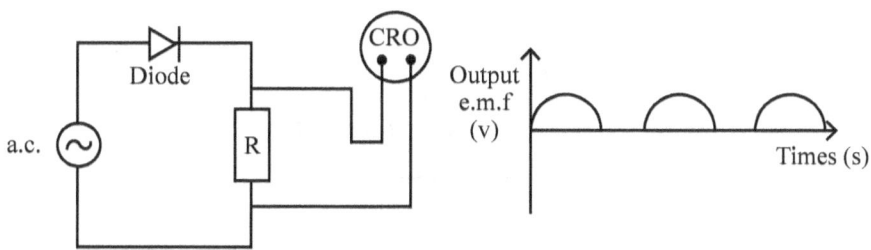

Smoothed output.

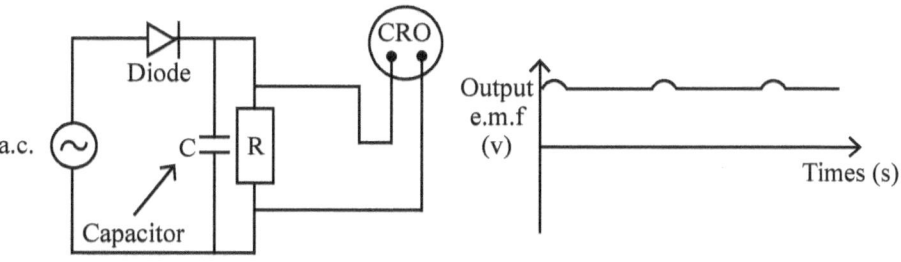

Faraday's law of electromagnetic induction states that the induced emf is directly proportional to the rate of change of magnetic flux and it opposes the motion causing it.

$$\varepsilon = -d(N\Phi)/dt$$

An induced emf also develops in an inductor (coil) when a current is passed through it. This is due to the self-inductance of the inductor L.
$$\varepsilon = -L\, dI/dt$$

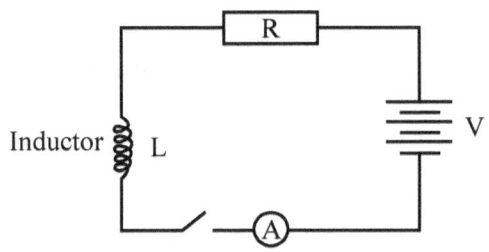

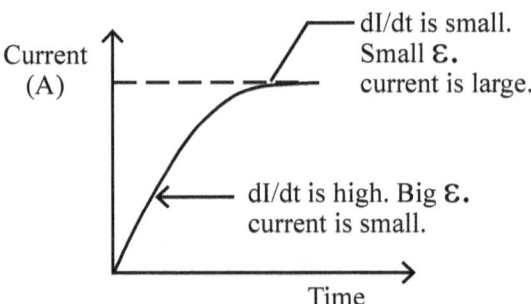

From the equation $L = -\varepsilon/dI/dt$
$$L = -\varepsilon dt/dI \ (VA^{-1}S)$$
Practical unit of self-inductance is Henry (H)
$$1H = 1VA^{-1}S.$$

Alternating Current

Concept Mapping

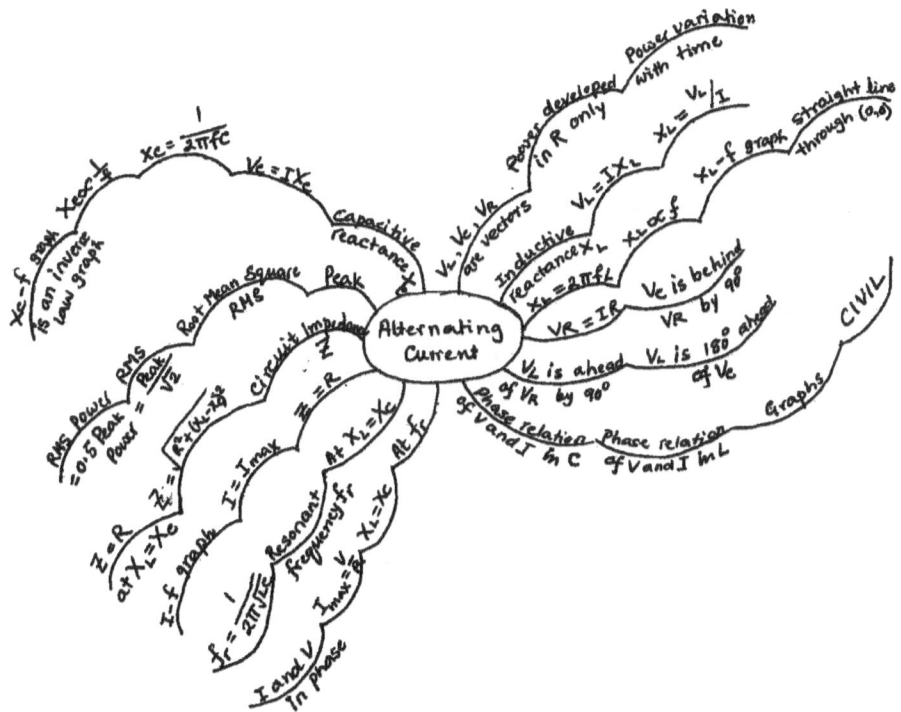

$X_L \Rightarrow$ Reactance of an inductor $= 2\pi f L$

$L \quad \Rightarrow$ Self-inductance of the inductor

$V_L \Rightarrow$ Voltage across the inductor $= IX_L = 2\pi I f L$

$X_C \Rightarrow$ Reactance of a capacitor $= 1/2\pi fC$

$C \quad \Rightarrow$ Capacitance of the capacitor

$Vc \Rightarrow$ Voltage across the capacitor $= IXc = I/(2\pi fC)$

$R \quad \Rightarrow$ Resistance of the resistor

$V_R \Rightarrow$ Voltage across the resistor $= IR$

X_L and X_C are frequency dependent quantities

R is not frequency dependent

V_L and V_C are frequency dependent quantities

V_R is not frequency dependent.

V_R, V_C, and V_L are vector quantities.

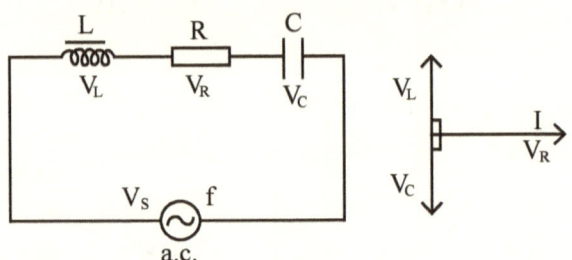

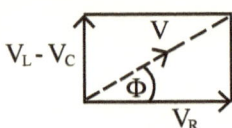

Resultant $V = \sqrt{(V_R^2 + (V_L - V_C)^2)}$

Impedance $Z = \sqrt{(R^2 + (X_L - X_C)^2)}$

$\phi = \tan^{-1}(X_L - X_C)/R$

Resonance occurs when $X_L = X_C$ and $Z = R$.

$$2\pi f_r L = 1/2\pi f_r C$$

Resonant frequency $f_r = 1/(2\pi\sqrt{(LC)})$.

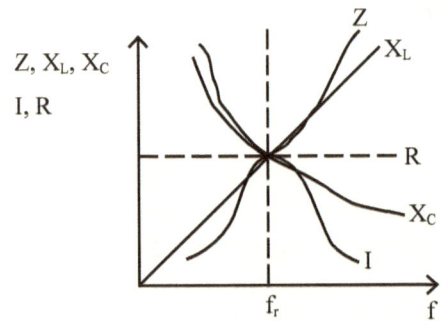

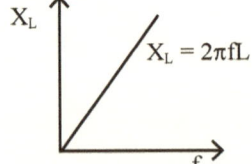

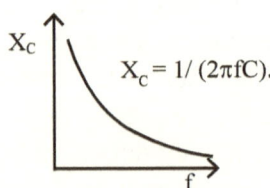

For an L–R circuit $Z = \sqrt{(R^2 + X_L^2)}$.

For a C–R circuit $Z = \sqrt{(R^2 + X_C^2)}$.

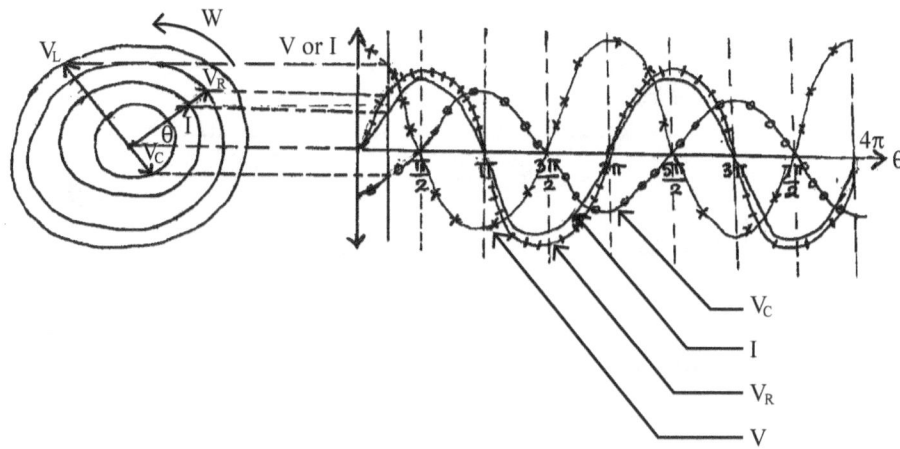

Phasor diagram to show the variation of V_L, V_R, I, and V_C in a circuit containing L, R, and C.

CIVIL to remember the phase relations
In C, I leads V by $\pi/2$.
In L, I lags V by $\pi/2$.
In R, V_R and I are in phase.

There is a phase of π between V_L and V_C.
The mean power developed in an L, RC circuit
    = RMS voltage × RMS current
    = $V_p/\sqrt{2} \times I_p/\sqrt{2}$ (P→peak value)
    = $I_p V_p/2$.

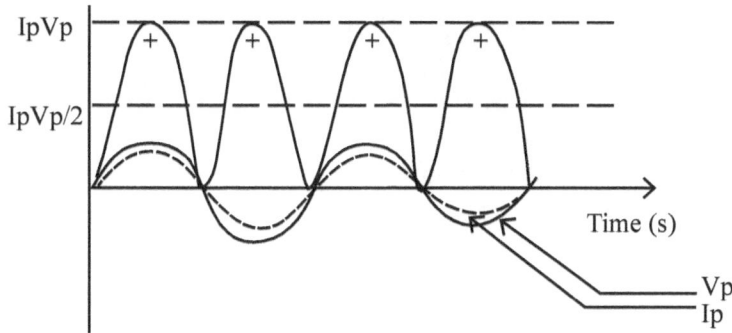

Power developed in R.
V_R and I are in phase.

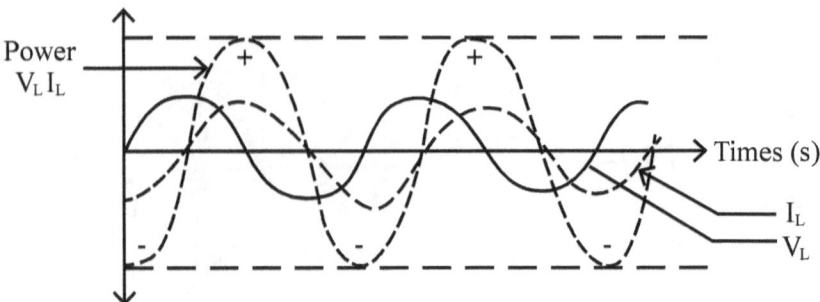

Power Developed in an Inductor

(+) Power is absorbed in the magnetic field of the inductor.
(–) Power is returned to the source of supply.
Therefore average power developed is zero.
No power is developed in a pure inductor.

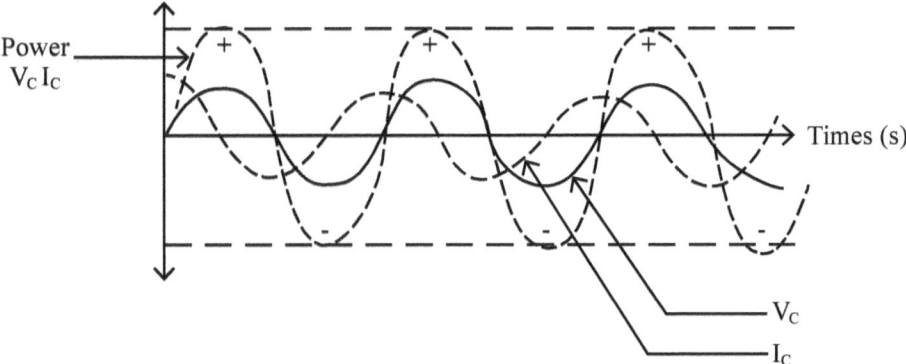

Power Developed in Capacitor

(+) Power is developed in the coulomb field of the capacitor as it is charging.
(–) Power is returned to the source of supply.
Therefore average power is zero.
No power is developed in a capacitor.

Section 15

Transfer Processes

Concept Mapping

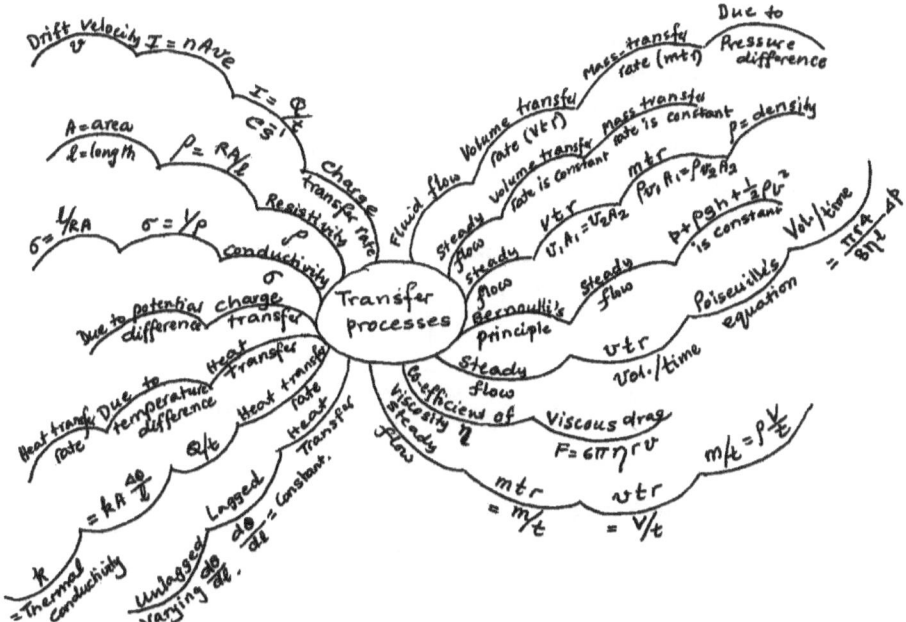

Transfer Processes

A steady-state fluid flow is due to pressure difference at two points.

Mass transfer rate m/t = ρ × (volume transfer rate V/t) where ρ = density.

At any two given points, volume transfer rate (vtr) $\Rightarrow V_1 A_1 = V_2 A_2$ and mass transfer rate (mtr) $\Rightarrow \rho V_1 A_1 = \rho V_2 A_2$.

Bernoulli's steady state equation: $P + \rho gh + 1/2\rho v^2$ = constant.

Poiseuille's volume transfer rate equation: vtr = $(\pi r^4 / 8\eta l) \Delta p$

where Δp = pressure difference.
Coefficient of viscosity for a steady-state laminar flow
    $\eta = (F/A)/ (v/l) = $ pressure/velocity gradient $= Fl/VA$
Viscous drag $F = 6\pi\eta rv$.

$$mtr = \rho(vtr) = (\rho\pi r^4/8\eta l)\, \Delta p$$
$$= \rho/8\pi\, (A2/\eta l)\, \Delta p$$
$$= C\, (A^2/\eta l)\, \Delta p. \quad (C = \rho/8\pi = \text{constant})$$

Process
  • Mass transfer due to pressure difference for a steady fluid flow
Formulae and conditions
  • Volume transfer rate is constant at any given point.
    $V_1 A_1 = V_2 A_2$ etc. (V $\Rightarrow$ velocity, A $\Rightarrow$ area)
  • Mass transfer rate is constant.
    $\rho V_1 A_1 = \rho V_2 A_2$ etc. (ρ = density of fluid)
  • Bernoulli's principle: $P + h\rho g + 1/2\rho v^2 = $ constant.
  • At a constant height, $P + 1/2\rho v^2 = $ constant.
  • From this equation, when V is big, P is small and vice versa.
Related formulae
  • Normal pressure $= F/A$ (Pa).
  • Fluid pressure $= h\rho g$ (Pa).

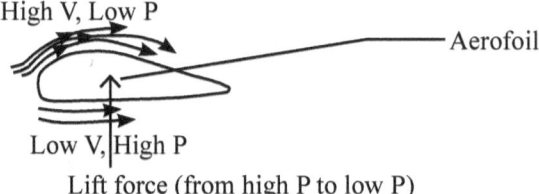

Lift force (from high P to low P)

Process
  • Charge transfer due to potential difference in a conductor.
Formulae and related conditions
  • $I = nAve$ (n = number of charge carriers per m^3, v = drift velocity)
  • Conductors have free electrons as charge carriers.
  • Electronic charge $e = 1.6 \times 10^{-19}$C.
  • Semiconductors have very few free electrons. In semiconductors, the charge carriers are electrons and holes. Whilst electrons have negative charge, holes have positive charge.
  • Insulators have no free electrons.

- In 1 m³ of copper, there are ≈ 10^{29} electrons.

Related formulae

- I = Q/t, R = ρl/A, ρ = RA/l, σ = 1/ρ, σ = l/RA (ρ = resistivity, σ = conductivity).
- In conductors, resistance increases with temperature.
- In semiconductors, resistance decreases with temperature.
- At very low temperatures, R = 0. Then we get superconductors used for making strong electromagnets.

Process

- Thermal energy transfer due to temperature difference in a conductor
- Thermal conduction

Formula and condition

- Rate of thermal energy transfer Q/t = kA($\theta 2$-$\theta 1$)/l (k = thermal conductivity)
- Free electrons are mostly responsible to carry thermal energy.
- Ionic vibrations in solids also contribute to thermal energy transfer by ion contacts.
- Thermal energy transfer occurs in solid conductors.

Related formulae

- Unit for thermal conductivity k is $Wm^{-1}K^{-1}$
- U-value = k/l, the unit is $Wm^{-2}k^{-1}$
- In lagged conductors, thermal energy flow is streamlined, and the temperature gradient is constant in the steady state.
- In un-lagged conductors, thermal energy flow is not streamlined, and the temperature gradient varies exponentially.

Lagged conductor.   *Un-lagged conductor.*

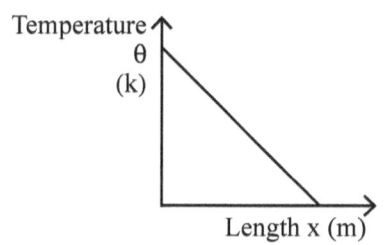

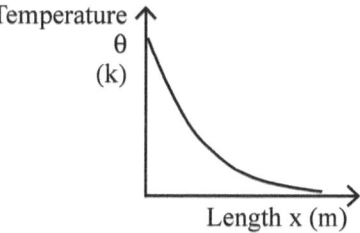

Process
- Thermal energy transfer due to temperature difference in a fluid

Thermal convection
- Formulae and conditions
- Natural convection in fluids is due to temperature difference
- Less-density fluid at higher temperature rises.
- High-density fluids at lower temperature falls.
- Convection current of fluid from high to low temperature is set up.
- When the process of convection is accelerated by an external force (e.g. blowing, using a fan etc.), it is called forced convection.

Related information
- Actual movement of fluid occurs.
- The process requires a fluid medium for the convection current to set up.

Process
Thermal energy transfer due to the temperature of the source, to the surroundings
Electromagnetic radiation
Energy transfer by electromagnetic wave
Formulae and conditions
The higher the temperature of the source, the greater is the power radiated.

Intensity Distribution

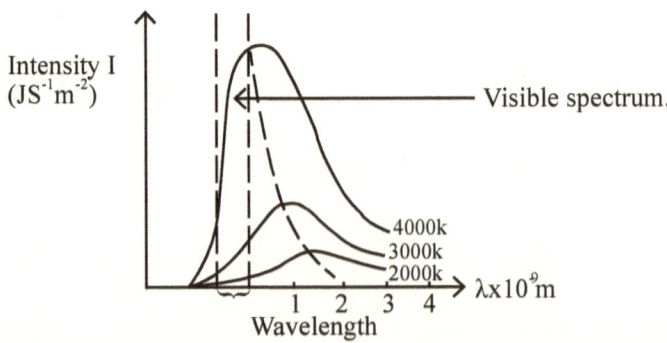

Related information
- Intensity I = power/area $\Rightarrow$ I = P/A (Wm^{-2} or Js^{-1}m^{-2}).
- Power = intensity $\times$ area.
- Spherical surface area = $4\pi r^2$.
- For a spherical source I = $P/4\pi r^2$.
- The area under the intensity distribution graph for a particular temperature gives the total radiant emittance of the hot body.
- This is found to be proportional to T^4.

Transfer Processes Compared

Mass flow rate
- m/t = C (A^2/ηl) Δp.

ηl/A^2 = resistance to mass flow rate.

Heat flow rate
- Q/t = (kA/l) $\Delta\theta$.

l/kA = resistance to heat flow.

Charge flow rate
- Q/t = (σA/l)ΔV.

l/σA = resistance to charge flow.

Section 16

Energy Changes and Gas Equations

Concept Mapping

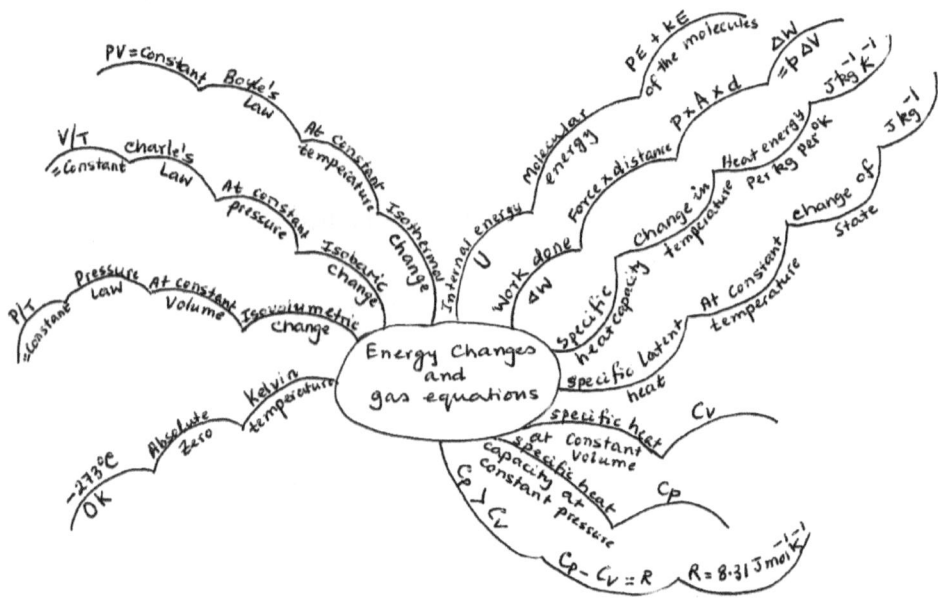

U = Internal energy	Sum of PE and KE of the molecules U can be changed by heating and/or working	$\Delta U = \Delta Q + \Delta W$ ΔU = Change in internal energy ΔQ = Change in heat (given) ΔW = Work done on the system.

| Work done ΔW | It is given by ΔW = pΔV
ΔV = change in volume | ΔW is positive when work is done on the system.
ΔW is negative when work is done by the system. |

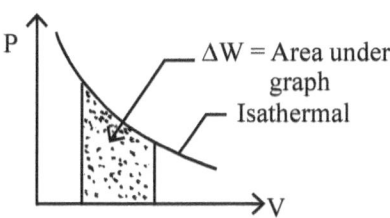

ΔW = Area under graph

Isathermal

| Specific heat capacity (change in temperature) | Heat energy required to change the temperature of 1kg of a substance by 1k. | $Q = mc\Delta\theta$.
C = specific heat capacity $(Jkg^{-1}k^{-1})$. |

| Specific latent heat (Change in state, no change in temperature). | Heat energy required for 1kg of a substance to change its state:
Solid ⇆ liquid or liquid⇆ gas. | $Q = \Delta ml$
l = specific latent heat (J kg^{-1}) |

| Specific heat capacity at constant volume for a gas (C_V). | Heat energy required to raise the temperature of 1kg of a gas by 1K when the volume is kept constant. | $C_p > C_V$, $C_p - C_V = R$
R = universal gas constant
= 8.31 $Jmol^{-1}k^{-1}$.
Here C_p and C_V are molar specific heats referred to 1mole of a gas. |

| Specific heat capacity at constant pressure for a gas (C_p). | Heat energy required to raise the temperature of 1kg of a gas by 1k when the pressure is kept constant. | |

Isothermal change (Temperature constant).	$P \propto 1/V \Rightarrow PV =$ constant $P_1 V_1 = P_2 V_2$	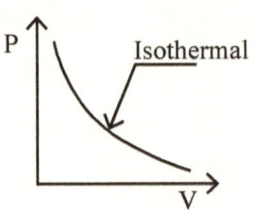
Isobaric change (Pressure constant).	$V \propto T \Rightarrow V/T =$ constant $V_1/T_1 = V_2/T_2$	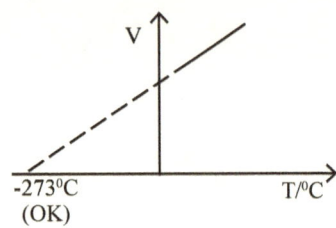
Isovolumetric change (Volume constant).	$P \propto T \Rightarrow P/T =$ constant $P_1/T_1 = P_2/T_2$	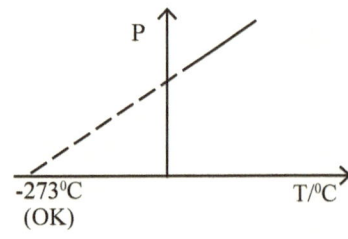
Adiabatic change.	The process in which there is no heating of the gas or by the gas. The system must be perfectly insulated from the surroundings.	$\Delta Q = 0.$
Work done in a complete cycle of change.	It is obtained from the area enclosed in a P-V diagram between two isothermals and two adiabatics.	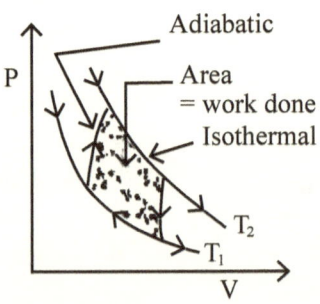

Gas pressure

Assumptions
- The gas is made up of similar particles
- There are N particles, each of mass m in a cube of volume l^3
- The particles collide with the walls perfectly elastically
- Newtonian mechanics is applied
- Volume of the particles is negligible compared with the volume of the gas as a whole.
- The pressure exerted on the walls is the resultant of all the pressure exerted on the walls by individual particles.

$\overline{\text{Derivation}}$ $P = (1/3)\rho c^2$

N = number of particles
n = number of moles
N_A = Avogadro number = 6.02×10^{23}
R = universal gas constant = 8.31 Jmol^{-1}K^{-1}.

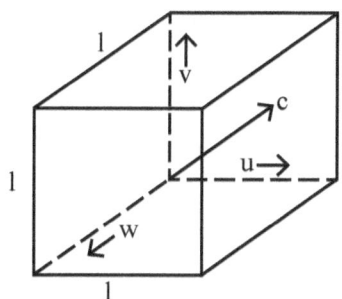

$N_A = N/n$
$k = R/N_A =$ Boltzmann constant
$k = 1.38 \times 10^{-23}$ JK^{-1}.

Δp = change in momentum = $2mu$ (reflecting on the wall)
$\Delta t = 2l/u$, $F = \Delta p/\Delta t = mu^2/l$, $P = F/l^2 = mu^2/l^3$
Total pressure = $m/l^3 \ (u^2_1 + u^2_2 + \cdots + u^2_N)$,

$\overline{u^2} = (u^2_1 + u^2_2 + \cdots + u^2_N) / N$

$P = Nm/l^3 \ \overline{u^2}$, if c = average speed of all the particles (rms),
r.m.s. $\Rightarrow$ root mean square

$\bar{c}^2 = \bar{u}^2 + \bar{v}^2 + \bar{w}^2 = 3\bar{u}^2 \ [u = v = w] \Rightarrow P = 1/3(Nm/l^3)\,\bar{c}^2 \Rightarrow P = 1/3\rho\bar{c}^2$
$\rho = Nm/l^3$

$PV = nRT$ and $PV = 1/3Nm\bar{c}^2$

$1/3\ Nm\bar{c}^2 = nRT$

$1/2\ m\bar{c}^2 \times 2/3N = nRT,\ 1/2m\bar{c}^2 = 3/2(n/N)\ RT = 3/2\ (R/N_A)\ T\ (N/n = N_A)$

$1/2\ m\bar{c}^2 = 3/2\ kT\ (k = R/N_A)$

Therefore, the mean KE of the gas $\propto$ T
At T = 0 mean KE = 0.

Section 17

Energy Levels and Line Spectra

Concept Mapping

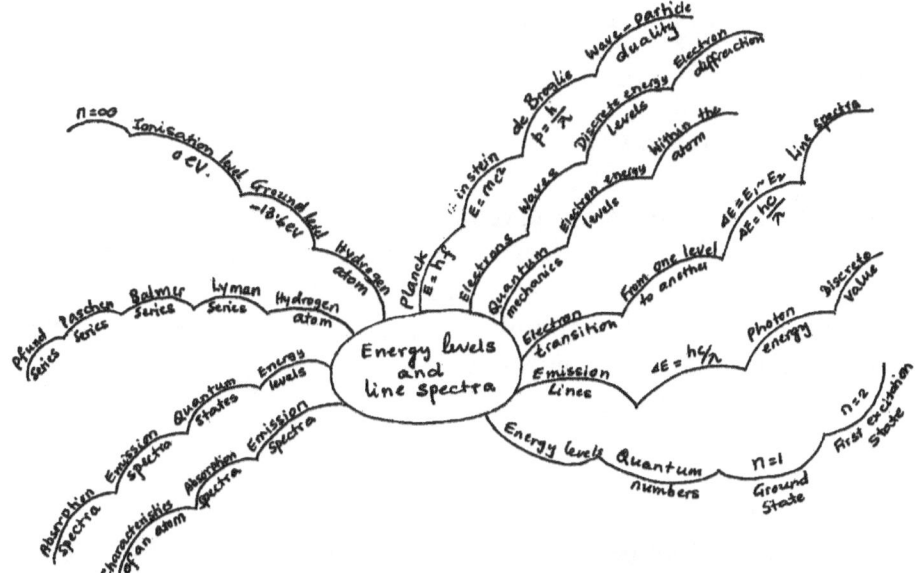

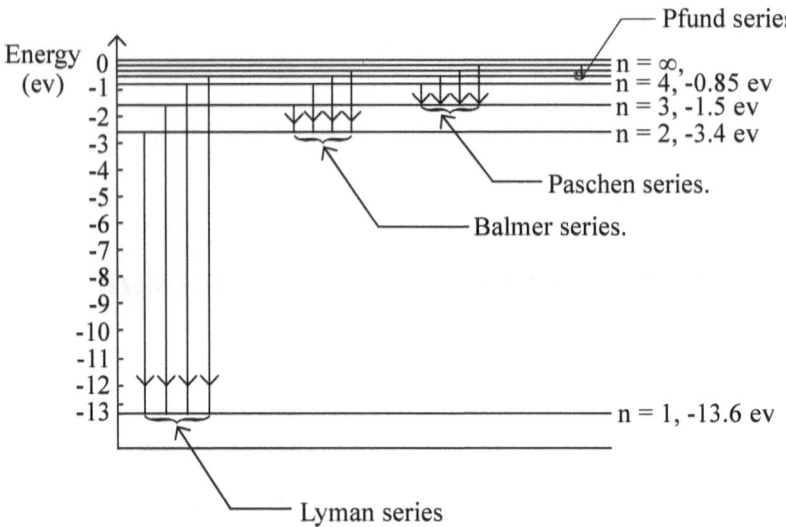

Electron energy levels of a hydrogen atom.

Spectrometers are used to observe line spectra of gases at a low pressure in a discharge tube. The observed line spectra are characteristic of the element. When a photon of light is emitted by an atom, the atom moves from a high-energy state to a low-energy state. The emitted wavelength of the photon can be obtained from:

$$\Delta E = E_1 \sim E_2$$
$$\Delta E = hc/\lambda$$

Electrons in the atom behave like a wave with a wavelength λ given by de Broglie's equation:

$$\lambda = h/p \ (h = \text{Planck's constant, } p = \text{momentum, } p = mc).$$

The electron can exist in the orbit around the nucleus if it has only whole number of wavelengths in the orbit. In other words, the electrons can have only discrete energy levels. Thus, energy associated with an electron is quantised.

In the energy level diagram for the hydrogen atom, $n = 1, 2, 3$, etc. are the quantum numbers corresponding to the energy levels.

n = 1, ground level ⇒ – 13.6 eV
n = 2, first excitation level ⇒ – 3.4 eV
n = 3, 2nd excitation level ⇒ – 1.5 eV
n = 4, 3rd excitation level ⇒ – 0.85 eV
n = 5, 4th excitation level ⇒ – 0.54 eV
n = ∞ ionisation level ⇒ zero energy.

Inside the atom, the energy values are negative. Beyond the ionisation level, the energy becomes positive. Removal of an electron from an atom increases the PE of the atom. Movement of an electron between energy levels results in emission and absorption spectrum.

Ionisation energy for hydrogen is 13.6 eV corresponding to a transition from n = 1 to n = ∞.

Transition from higher levels to n = 1, gives the Lyman series.
Transition from higher levels to n = 2, gives the Balmer series.
Transition from higher levels to n = 3, gives the Paschen series.
Transition from higher levels to n = 4, gives Brackett series.
Transition from higher levels to n = 5, gives Pfund series.
The shortest wavelength in the Lyman series is given by

$$\lambda_{min} = hc/\Delta E$$
$$= 6.6 \times 10^{-34} \times 3 \times 10^{8}/13.6 \times 1.6 \times 10^{-19}$$
$$= 9.1 \times 10^{-8} m.$$

An external energy input to the atom leaves the atom in an excited state. An electron from the stable ground state moves to an excited state. The electron then has an extra energy, which will be released as a quantum of energy (E = hf) when the electron jumps back to the ground state at a subsequent stage. This appears as a line spectrum when viewed through a spectrometer.

This can be illustrated by a simple model of virtual quantum jump of a sandpiper 'Twitee'.

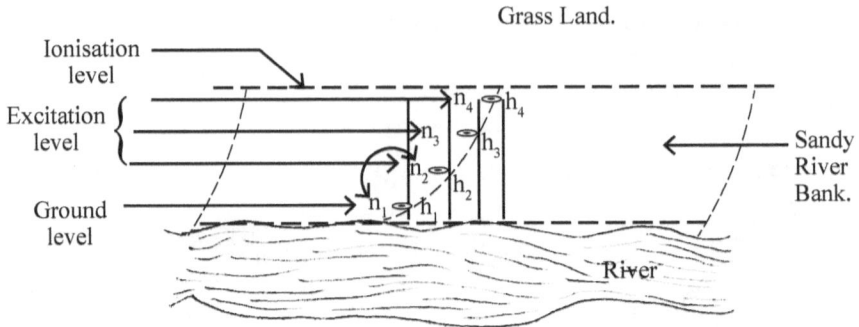

Twitee has four nests, n_1 n_2, n_3 and n_4, built in sand. Every time the river swells, she goes to the next nest higher up. Her stable nest is n_1, where she is most comfortable. When she moves to the nests at higher levels (for n_2, n_3, n_4), she gains PE. When the water level goes down, Twitee can go down to lower levels gaining KE. If the water level rises to flood the whole sandy bank, our Twitee will have to fly high above the grassland up to the tree (i.e. ∞). She is then in an ionised state. Whilst n_1 corresponds to the ground state, n_2, n_3, and n_4 represent the excited states. The ionisation level is the grassland at the top of the sandy bank.

Section 18

Nucleonics

Concept Mapping

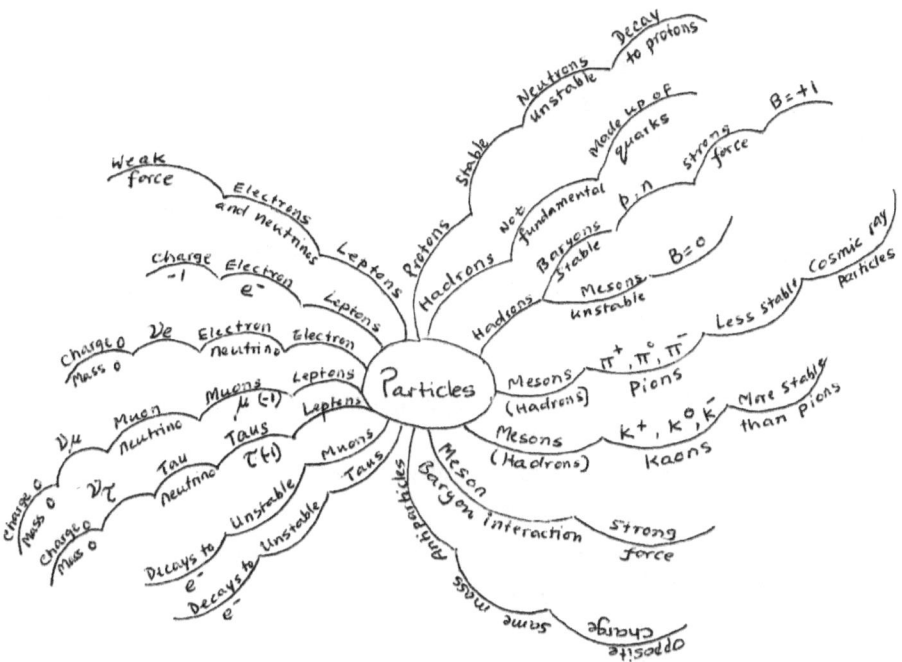

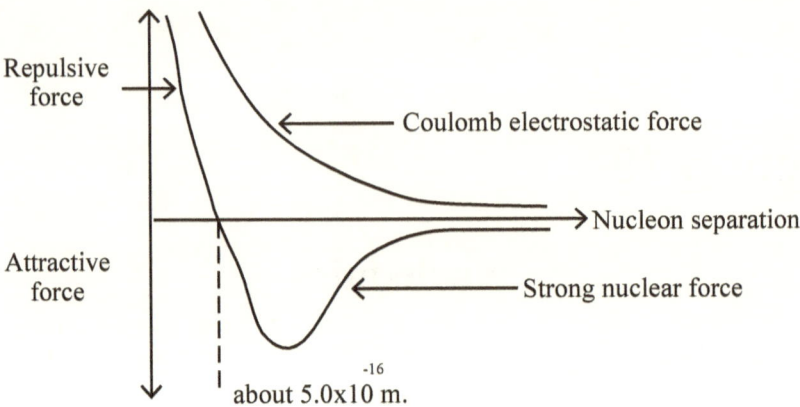

about 5.0×10^{-16} m.

The graph shows the forces involved in holding the nucleons together in a nucleus. Coulomb electrostatic force is a repulsive force which extends to a large nucleon separation obeying inverse-square law. The strong nuclear force is a repulsive force upto about 5.0×10^{-16}m, beyond which it becomes an attractive force, reaches a maximum attractive value, and then drops quickly to very small values. This explains why at a nuclear radius of about 1×10^{-10}m, the nucleons are held together without getting fused.

Protons and neutrons are held together by the strong nuclear force.
Protons and neutrons are hadrons.
Hadrons have two types: baryons and mesons.
Baryons $\Rightarrow$ protons, neutrons, and Σ
All baryons decay to protons. Hence, protons are stable.
Protons and neutrons $\Rightarrow$ baryon number $B = +1$.

Mesons have $B = 0$ and are unstable.
π^+, $\pi-$ and π° mesons have the smallest mass.
K^+, $K-$ and K° mesons are heavier and more unstable.
Meson–baryon interaction involves strong force.

Leptons $\Rightarrow$ electrons (e−), muons (μ−), and tau (τ−).
Leptons take part in weak interactions.
Electrons are stable; muons and taus are unstable.
Muons and taus decay to electrons.
These three particles have their corresponding neutrinos v_e, v_μ, and v_τ.

Neutrinos have zero charge and zero mass.
Neutrinos take part in weak interactions.
Leptons are fundamental particles.
Electrons, muons, and taus have a charge of -1.
v_e, v_μ and v_τ have zero charge.
There are three different Lepton numbers L_e, L_μ, and L_τ.

	L_e	L_μ	L_τ
e−	+1	0	0
v_e	+1	0	0
μ^-	0	+1	0
v_μ	0	+1	0
τ−	0	0	+1
v_τ	0	0	+1

Quarks: Building Blocks of Hadrons

Concept Mapping

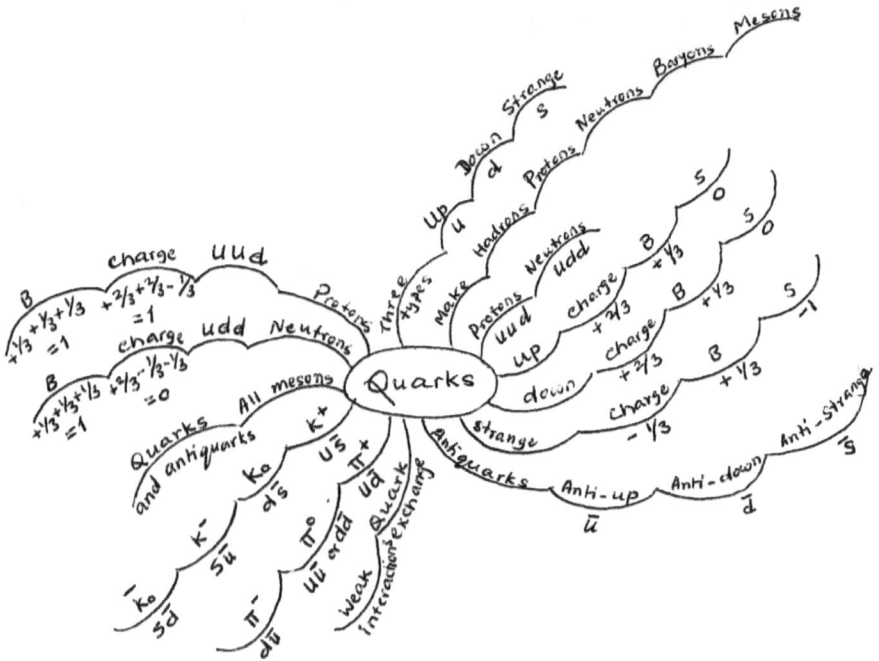

Since the discovery of positron, which is an electron with at positive charge, every particle has been found to have an antiparticle. The antiparticles have the same mass as the particle but have opposite charge.

If a gamma photon has enough energy, it can produce an electron–positron pair. This is called pair production. This is based on Einstein's mass–energy equivalence equation

$$E = mc^2.$$

The photon energy has been converted into two particles $\Rightarrow$ electron and positron, travelling at the speed of light.

If the particle and the antiparticle collide, the total mass is converted into energy. Electron–positron collision results in the production of two γ-ray photons. This is called annihilation.

Quarks are fundamental particles.
Hadrons (baryons and mesons) are built up of quarks.
Quarks are up, down, and strange (u, d, and s respectively).
Quarks have their antiquarks $\Rightarrow$ anti-up ($\bar{u}$), anti-down ($\bar{d}$) and anti-strange ($\bar{s}$).
Quark structure of proton is uud.
Quark structure of neutron is udd.
Mesons have two quarks $\Rightarrow$ quark and an antiquark.

	Charge Q	Baryon no. B	Strangeness S
u	+2/3	+1/3	0
d	−1/3	+1/3	0
s	−1/3	+1/3	−1

Antiquarks have opposite properties to those of quarks.
For a proton (uud), Q = 2/3 + 2/3 − 1/3 = 1, B = 1/3 + 1/3 + 1/3 = 1.
For a neutron (udd), Q = 2/3 − 1/3 − 1/3 = 0, B = 1/3 + 1/3 + 1/3 = 1

Quark exchange takes place in a weak interaction. In the β-decay of neutron, the neutron changes into a proton. A d quark changes into a u quark in the process.

n (udd) $\rightarrow$ p (uud) + electron (e-) + antineutrino

In a particle interaction

- Charges are conserved.
- Baryon numbers are conserved.
- Strangeness is conserved (in strong interactions).

- All lepton numbers Le, Lμ, and Lτ must be counted for the reacting particles and the product particles to show conservation.

Particle Links

Learning Zone

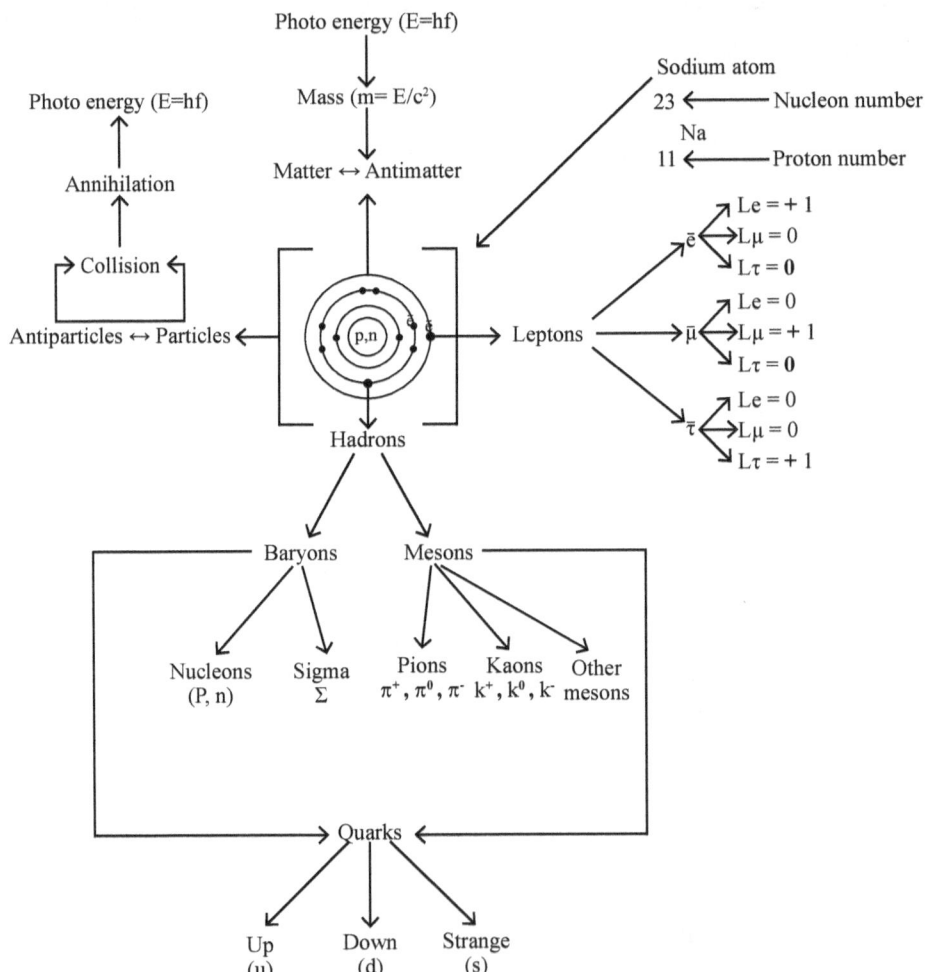

INDEX

www.ingramcontent.com/pod-product-compliance
Lightning Source LLC
Chambersburg PA
CBHW032012170526
45157CB00002B/667